Health Impact Assessment of

SHALE GAS EXTRACTION

WORKSHOP SUMMARY

Christine Coussens and Rose Marie Martinez, *Rapporteurs*

Roundtable on Environmental Health Sciences, Research, and Medicine

Board on Population Health and Public Health Practice

INSTITUTE OF MEDICINE
OF THE NATIONAL ACADEMIES

THE NATIONAL ACADEMIES PRESS
Washington, D.C.
www.nap.edu

THE NATIONAL ACADEMIES PRESS 500 Fifth Street, NW Washington, DC 20001

NOTICE: The workshop that is the subject of this workshop summary was approved by the Governing Board of the National Research Council, whose members are drawn from the councils of the National Academy of Sciences, the National Academy of Engineering, and the Institute of Medicine.

This activity was supported by contracts between the National Academy of Sciences and the National Institute of Environmental Health Sciences, The Kresge Foundation, American Chemistry Council, Colgate-Palmolive Company, ExxonMobil Foundation, and Royal Dutch Shell. The views presented in this publication do not necessarily reflect the views of the organizations or agencies that provided support for the activity.

This summary is based on the proceedings of a workshop that was sponsored by the Roundtable on Environmental Health Sciences, Research, and Medicine. It is prepared in the form of a workshop summary by and in the name of the rapporteurs as an individually authored document.

International Standard Book Number-13: 978-0-309-28791-3
International Standard Book Number-10: 0-309-28791-X

Additional copies of this report are available from the National Academies Press, 500 Fifth Street, NW, Keck 360, Washington, DC 20001; (800) 624-6242 or (202) 334-3313; http://www.nap.edu.

For more information about the Institute of Medicine, visit the IOM home page at: **www.iom.edu.**

Printed in the United States of America

The serpent has been a symbol of long life, healing, and knowledge among almost all cultures and religions since the beginning of recorded history. The serpent adopted as a logotype by the Institute of Medicine is a relief carving from ancient Greece, now held by the Staatliche Museen in Berlin.

Suggested citation: IOM (Institute of Medicine). 2014. *Health impact assessment of shale gas extraction: Workshop summary.* Washington, DC: The National Academies Press.

"Knowing is not enough; we must apply.
Willing is not enough; we must do."
—Goethe

INSTITUTE OF MEDICINE
OF THE NATIONAL ACADEMIES

Advising the Nation. Improving Health.

THE NATIONAL ACADEMIES

Advisers to the Nation on Science, Engineering, and Medicine

The **National Academy of Sciences** is a private, nonprofit, self-perpetuating society of distinguished scholars engaged in scientific and engineering research, dedicated to the furtherance of science and technology and to their use for the general welfare. Upon the authority of the charter granted to it by the Congress in 1863, the Academy has a mandate that requires it to advise the federal government on scientific and technical matters. Dr. Ralph J. Cicerone is president of the National Academy of Sciences.

The **National Academy of Engineering** was established in 1964, under the charter of the National Academy of Sciences, as a parallel organization of outstanding engineers. It is autonomous in its administration and in the selection of its members, sharing with the National Academy of Sciences the responsibility for advising the federal government. The National Academy of Engineering also sponsors engineering programs aimed at meeting national needs, encourages education and research, and recognizes the superior achievements of engineers. Dr. C. D. Mote, Jr., is president of the National Academy of Engineering.

The **Institute of Medicine** was established in 1970 by the National Academy of Sciences to secure the services of eminent members of appropriate professions in the examination of policy matters pertaining to the health of the public. The Institute acts under the responsibility given to the National Academy of Sciences by its congressional charter to be an adviser to the federal government and, upon its own initiative, to identify issues of medical care, research, and education. Dr. Harvey V. Fineberg is president of the Institute of Medicine.

The **National Research Council** was organized by the National Academy of Sciences in 1916 to associate the broad community of science and technology with the Academy's purposes of furthering knowledge and advising the federal government. Functioning in accordance with general policies determined by the Academy, the Council has become the principal operating agency of both the National Academy of Sciences and the National Academy of Engineering in providing services to the government, the public, and the scientific and engineering communities. The Council is administered jointly by both Academies and the Institute of Medicine. Dr. Ralph J. Cicerone and Dr. C. D. Mote, Jr., are chair and vice chair, respectively, of the National Research Council.

www.national-academies.org

PLANNING COMMITTEE FOR THE WORKSHOP ON THE HEALTH IMPACT ASSESSMENT OF NEW ENERGY SOURCES: SHALE GAS EXTRACTION[1]

ROB DONNELLY, Royal Dutch Shell, The Hague, Netherlands
LYNN R. GOLDMAN, George Washington University, Washington, DC
GEORGE M. GRAY, George Washington University, Washington, DC
ANDREW MAGUIRE, Environmental Defense Fund, Washington, DC
LINDA A. McCAULEY, Emory University, Atlanta, GA
AUBREY MILLER, National Institute of Environmental Health Sciences, Bethesda, MD
CHRISTOPHER J. PORTIER, Centers for Disease Control and Prevention, Atlanta, GA
SCOTT W. TINKER, University of Texas, Austin

[1] Institute of Medicine planning committees are solely responsible for organizing the workshop, identifying topics, and choosing speakers. The responsibility for the published workshop summary rests with the workshop rapporteurs and the institution.

ROUNDTABLE ON ENVIRONMENTAL HEALTH SCIENCES, RESEARCH, AND MEDICINE[1]

[1] Institute of Medicine forums and roundtables do not issue, review, or approve individual documents. The responsibility for the published workshop summary rests with the workshop rapporteurs and the institution.

JOHN D. SPENGLER, Harvard School of Public Health, Boston, MA
LOUIS W. SULLIVAN, Morehouse School of Medicine, Atlanta, GA
ANNE M. SWEENEY, Texas A&M University, College Station
G. DAVID TILMAN, University of Minnesota, St. Paul
PATRICIA VERDUIN, Colgate-Palmolive Company, Piscataway, NJ
NSEDU OBOT WITHERSPOON, Children's Environmental Health Network, Washington, DC
HAROLD ZENICK, U.S. Environmental Protection Agency, Research Triangle Park, NC

IOM Staff

CHRISTINE COUSSENS, Study Director (until August 2013)
ERIN RUSCH, Associate Program Officer
ANDRÉS GAVIRIA, Research Associate
SUZANNE LANDI, Research Associate
ANDREW LEMERISE, Research Associate
HOPE HARE, Administrative Assistant
ROSE MARIE MARTINEZ, Director, Board on Population Health and Public Health Practice

Reviewers

This workshop summary has been reviewed in draft form by individuals chosen for their diverse perspectives and technical expertise, in accordance with procedures approved by the National Research Council's Report Review Committee. The purpose of this independent review is to provide candid and critical comments that will assist the institution in making its published workshop summary as sound as possible and to ensure that the workshop summary meets institutional standards for objectivity, evidence, and responsiveness to the study charge. The review comments and draft manuscript remain confidential to protect the integrity of the process. We wish to thank the following individuals for their review of this workshop summary:

Jerome Paulson, Mid-Atlantic Center for Children's Health and the Environment
Susan Santos, Rutgers School of Public Health
Kyra Naumoff Shields, University of Pittsburgh
Leonardo Trasande, New York University

Although the reviewers listed above have provided many constructive comments and suggestions, they did not see the final draft of the workshop summary before its release. The review of this workshop summary was overseen by **Melvin Worth.** Appointed by the Institute of Medicine, he was responsible for making certain that an independent examination of this workshop summary was carried out in accordance with institutional procedures and that all review comments were carefully considered. Responsibility for the final content of this workshop summary rests entirely with the rapporteurs and the institution.

Contents

1

Introduction

Natural gas extraction from shale formations—which includes hydraulic fracturing—is increasingly in the news as the use of extraction technologies has expanded, rural communities have been transformed seemingly overnight, public awareness has increased, and regulations have been developed. The governmental public health system, which retains primary responsibility for health, was not an early participant in discussions about shale gas extraction; thus public health is lacking critical information about environmental health impacts of these technologies and is limited in its ability to address concerns raised by regulators at the federal and state levels, communities, and workers employed in the shale gas extraction industry.

In public health, there is an increasing awareness of the importance of a "health-in-all" policy approach to protecting human health (IOM, 2011). Central to this approach is the use of health impact assessments to inform decisions by providing a structured process that uses scientific data, professional expertise, and stakeholder input to identify and evaluate the public health consequences of policy and program proposals. In 2011 the National Research Council (NRC) published a report, *Improving Health in the United States: The Role of Health Impact Assessment,* which provides recommendations on how to conduct health impact assessments and discusses how these can be used to minimize adverse health effects and optimize the beneficial effects of the policy or program assessed (NRC, 2011).

Members of the Institute of Medicine (IOM) Roundtable on Environmental Health Sciences, Research, and Medicine recognized the need to discuss the human health impact of shale gas extraction through the lens of a health impact assessment; to that end, it convened a public workshop in Washington, DC, in 2012. Through invited presentations and discussions, participants examined the state of the science regarding shale gas extraction, the direct and indirect environmental health impacts of shale gas extraction, and the use of health impact assessment as a tool that can help decision makers identify the public health consequences of shale gas extraction. The statement of task for the workshop can be found in Box 1-1.

The workshop was organized by an independent planning committee in accordance with the procedures of the NRC. The planning group included Rob Donnelly, Lynn Goldman, George Gray, Andrew Maguire, Linda McCauley, Aubrey Miller, Christopher Portier, and Scott Tinker; their role was limited to planning the workshop. They developed the agenda topics and selected and invited expert speakers and discussants to address identified topics. This summary was prepared by the workshop rapporteurs as a factual summary of what occurred at the workshop. All views presented in the summary are those of the individual workshop participants. The summary does not contain any findings or recommendations by the planning committee or the Roundtable.

The workshop was moderated by roundtable members and featured presentations and discussion. Chapter 2 presents a summary of opening remarks and two opening presentations—one that frames the objectives of the workshop and one that describes health impact assessment. Chapter 3 provides a summary of presentations that describe the process of shale gas extraction and the geographic footprint or changes that occur in the environment. Chapter 4 summarizes occupational hazards associated with shale gas extraction and potential impacts on communities. Chapters 5 and 6 summarize presentations on the impact of shale gas extraction on air quality and water resources, respectively. Chapter 7 summarizes presentations on the broad topic of sustainable energy options and the need to ensure health in all approaches. Chapters 8 and 9 present perspectives from those in the research community and federal agency representatives, respectively, regarding research gaps and opportunities. Highlights from the concluding discussion are provided at the end of Chapter 9.

BOX 1-1
Statement of Task

An ad hoc committee will plan and conduct a public workshop on the human health impacts of shale gas extraction. The workshop will feature invited presentations and discussions to look at the state of the science in shale gas extraction, direct and indirect environmental health impacts of shale gas extraction, and the role of health impact assessments in minimizing health impacts. The committee will develop the workshop agenda, select invited speakers and discussants, and moderate the discussions. A workshop summary will be prepared by a designated rapporteur in accordance with National Research Council policies and procedures.

REFERENCES

IOM (Institute of Medicine). 2011. *For the public's health: Revitalizing law and policy to meet new challenges*. Washington, DC: The National Academies Press.

NRC (National Research Council). 2011. *Improving health in the United States: The role of health impact assessment.* Washington, DC: The National Academies Press.

2

Opening Session

This chapter provides a summary of the opening remarks and presentations that framed the workshop. Presentations discussed the growing use of hydraulic fracturing to extract natural gas from shale rock, and the health and other related concerns raised by community members. The Centers for Disease Control and Prevention (CDC) programs that support the assessment of community concerns and that provide monitoring of potential adverse health impacts are briefly described. Health impact assessment (HIA) is discussed as a potential systematic process (based on data, analysis, and input from stakeholders) to determine the potential effects that a proposed project may have on the health of a population. HIA is used to inform decision making and create opportunities for planning and management of the potential negative impact of a project.

OPENING REMARKS

Harvey V. Fineberg, M.D.
President, Institute of Medicine

Harvey V. Fineberg opened the workshop by explaining that the Institute of Medicine (IOM) is the health arm of the National Academy of Sciences. The National Academy of Sciences, together with the National Academy of Engineering, the National Research Council, and the IOM constitute the National Academies. The IOM, he noted, like its sister academies, is an honorific association. It was chartered in 1970, but its mission is to serve as an advisor to the nation, to government, to the citizens, and to interested parties on matters related to health. The mission of the IOM is to provide unbiased, evidence-based, and authoritative information and recommendations related to health and science policy, to policy makers, to health professionals, to interested groups, to businesses, and to the citizens at large. The IOM also serves as a convener of thought leaders and experts on matters of science and policy—these occasions are intended as regular opportunities to have a dialogue on important health challenges.

Dr. Fineberg further noted that the IOM uses roundtables as an opportunity for all interested parties (i.e., government, industry, interested groups, and representatives of citizens at large) to gather periodically to engage in free exchange. The Roundtable on Environmental Health Sciences, Research, and Medicine, the sponsor of the workshop, has adopted and emphasized a theme of health-in-all policy. That approach, he noted, recognizes that many aspects of decision making and choice in society, whether it is about housing, education, our transportation system, or certainly about our environment, all have potential and documented impacts on human health. Thus, the Roundtable is interested not only in learning about the sources of these influences on health, but also in exploring options and opportunities to improve health through improved policies in all of these areas.

The workshop, he stated, brings together many eminent scientists, physicians, public health experts, and representatives from government agencies at federal and state levels, from nongovernment organizations, from the business sector, and from interest groups representing the interests of the citizens to exchange ideas and to inform on a very challenging problem—hydraulic fracturing as a means of extraction of natural gas. It is a challenging question that has many dimensions to be illuminated in order to understand more clearly what is at stake and what can be done.

INTRODUCING THE WORKSHOP THEME

Lynn R. Goldman, M.D.
Vice Chair, Roundtable on Environmental Health Sciences, Research, and Medicine
Dean, George Washington University School of Public Health and Health Services

Lynn R. Goldman began her remarks by highlighting that when first convened in 1998 the Roundtable adopted a broad view of environmental health—one that includes the natural, built, and social environments and considers how changes in the environment can impact human health (IOM, 2006). The roundtable has used a life-cycle approach that includes both upstream and downstream drivers to understand health impacts. For example, for a product on the market there may be health implications during production, use, and disposal that need to be taken into account. Because environmental protection is also health protection, the Roundtable has been drawn toward a number of issues, including energy.

Dr. Goldman noted that energy is today a predominant theme in national policy discussions. There are debates about drilling along the Gulf Coast, building the Keystone Pipeline, and opportunities for U.S. energy independence. She stated that over the next few years and decades, the United States and many other countries will be making important decisions about where to invest resources for ensuring that energy needs are being met. These decisions will not be easy to make; there is no clear path forward and trade-offs will need to be made. She emphasized that policy decisions often focus on energy needs, environmental impact, or economic considerations. In many of these decisions, the health community is not at the table. She stated that health impacts of energy alternatives are important and need to be evaluated and discussed early in the decision-making processes. When health impacts are considered late in the process after decisions have been made, then needed adjustments to protect health too often are difficult and expensive. When considered in a timely matter, health decisions do not have to be at odds with economic growth, but can serve as a partner in ensuring an energy future.

During the planning of the workshop the planning committee recognized that the discussion on shale gas extraction and energy production in general is part of a much larger and more complex picture of a major societal transformation of how we generate and use energy for the power grid and for transportation. As the country begins to think about ensuring sustainable energy, there will be a need to continue these discussions in order to have a more complete view.

Members of the Roundtable have heard concerns and anecdotal reports of possible health problems related to shale gas extraction from communities and have read some of the preliminary research in this area. The Roundtable will not be making any conclusions or recommendations, but rather this workshop gathers experts in the field to better understand the state of the science as it exists in April 2012. Dr. Goldman emphasized that comments made by individuals, including members of the Roundtable, should not be interpreted as positions of the Roundtable, the IOM, or its sponsors.

FRAMING THE WORKSHOP

Christopher J. Portier, Ph.D.
Director of the National Center for Environmental Health and Agency for Toxic Substances and Disease Registry
Centers for Disease Control and Prevention

Christopher J. Portier began his presentation by noting that improved methods for recovering natural gas from large reserves in the United

States have emerged in recent years. The expansion of natural gas extraction has led to new jobs and improved economies in states and communities across the country. If lessons from past experiences can be drawn, over time these improved economies will lead to improved public health in these communities. But improved economies can also introduce new public health concerns. He noted that some reports to the Agency for Toxic Substance and Disease Registry (ATSDR) have raised concerns about the potential for adverse health effects on individuals and communities involved in hydraulic fracturing.

Dr. Portier stated that most notable among these concerns are: the potential for chemical contamination of drinking water wells from hydraulic fracturing fluids, the mobilization of naturally occurring metals and salts, the potential for emission of volatile organics into the air, the potential for explosive hazards from buildup of methane in drinking water wells, increased traffic leading to reductions in road safety, the potential for stress to the medical and emergency response systems and infrastructure in communities, and changes in the overall quality of life that could adversely affect human health. He emphasized that although the CDC and ATSDR are not regulatory agencies, they serve a critical role in protecting environmental public health in this nation by protecting the nation from immediate public health threats 24 hours a day and through the development of programs that prevent disease. Prevention not only saves lives but it can save money through reducing or eliminating health care costs and losses in productivity, he stated.

Dr. Portier proceeded to outline programs that address community concerns related to shale gas extraction under the National Center for Environmental Health (NCEH) and ATSDR at the CDC. ATSDR works in communities, evaluating the potential for environmental exposures that can negatively affect human health. The agency has received calls from community members, community organizations, media, and local and state governments. These callers have raised a number of health concerns related to shale gas extraction, including nausea, respiratory issues, and irritating odors. Further, individuals living near drilling sites have also expressed concern about potential long-term health effects. The agency is working to address these community concerns.

NCEH also addresses community concerns through its Environmental Public Health Tracking Network and the National Biomonitoring Program. The Environmental Public Health Tracking Network[1] tracks both exposures and disease incidence and prevalence at the county level to understand trends in the environment–disease linkage and to collect evidence on the effectiveness of interventions. The National Biomonitoring Program[2] has the capability to measure more than 450 chemicals and nutrition indicators in human tissues and urine to understand the

---

[1] See http://ephtracking.cdc.gov/showHome.action (accessed May 30, 2013).
[2] See http://www.cdc.gov/biomonitoring (accessed May 30, 2013).

magnitude of exposures at the individual level (CDC, 2012). Additionally, ATSDR and NCEH are supporting other scientific efforts by providing public health and toxicological expertise to the U.S. Environmental Protection Agency's (EPA's) National Hydraulic Fracturing Study.[3]

Dr. Portier then proceeded to describe the focus of the workshop. He noted that the intersection of health and hydraulic fracturing is a rapidly evolving issue. Currently, 16 states have begun or are contemplating shale production, which underscores the importance of this topic for many individuals across the United States. He said that the objective of the workshop is, through scientific discussion, to shed light on whether shale gas extraction poses potential public health challenges and the extent of the concerns for the nation.

Dr. Portier further commented that on January 13, 2012, the White House released a report of the administration's support for economic investments, including support for natural gas development (White House, 2012). He also noted that President Obama has stated that America will develop this resource without putting the health and safety of our citizens at risk (Executive Order 13605, 2012). Public health professionals recognize that health is positively linked to job growth and economic development. However, public health officials are also committed to ensuring that economic development progresses responsibly and in a way that addresses the health and safety concerns of Americans. Thus, the steps taken now to identify health threats can be used to improve shale gas extraction techniques, allowing all to continue to advance the economic and security goals of this nation while protecting health and preventing disease.

HEALTH IMPACT ASSESSMENT FOR SHALE GAS EXTRACTION

Aaron Wernham, M.D., M.S.
Project Director, Health Impact Project
Pew Charitable Trusts

Aaron Wernham began his presentation by noting that the use of HIA is relatively new in the United States and has been growing in use over the past 10–12 years as a mechanism to conduct health-oriented planning and decision making. HIA has been used primarily in natural resource development, and only recently has there been interest in applying principles to other areas of decision making such as hydraulic fracturing.

---

[3] See http://www2.epa.gov/hfstudy (accessed May 30, 2013).

Dr. Wernham explained that HIA is

> a systematic process that uses an array of data sources and analytic methods and considers input from stakeholders to determine the potential effects of a proposed policy, plan, program, or project on the health of a population and the distribution of those effects within the population. HIA provides recommendations on monitoring and managing those effects. (NRC, 2011)

HIA is not a health risk assessment. HIA is a management tool to assess complex societal decisions that may have health implications and options for managing the health effects. It is not meant to just identify risks, and its purpose is not to determine if a proposal or policy is a good idea or not. HIA

- informs decision making on a specific proposed action (legislation, new regulation, permit, growth plan, and so forth);
- identifies a broad range of potential risks and benefits of the proposal;
- emphasizes interagency collaboration;
- solicits input from stakeholders (regulators, industry, community, and so forth); and
- offers recommendations to address data gaps, establish a monitoring framework, maximize benefits, and minimize any risks.

HIA is primarily qualitative in nature and does not make quantitative comparisons across choices. The focus is to identify the different ways, both direct and indirect, that health may be affected. Further, HIA is more than just an analytic tool; it is an opportunity for dialogue. During the development of an HIA, the process provides an opportunity for stakeholders to have a reasoned discussion about what they view are the risks and how to manage them (NRC, 2011).

There are six steps to develop HIAs:

1. Screening. Decide whether HIA should be done; consider if it will add information and can be done within the time line.
2. Scoping. Develop the framework for the HIA; identify the most important health effects and affected populations to assess with the available evidence.
3. Assessment. Analyze the baseline conditions or characteristics of the population and predict potential effects.
4. Recommendations. Develop health-based recommendations, a feasible plan for implementing them, and indicators for monitoring.
5. Reporting. Develop a report, disseminate the results to decision makers, the public, and other stakeholders.

6. Monitoring and Evaluation of the HIA Process. Determine whether it added value to the decision-making process; evaluate the outcomes of implementing HIA recommendations. (NRC, 2011)

Although HIA is a growing field in the United States, the international community, including the World Bank and the International Finance Corporation (IFC), routinely uses this tool. HIA in international practice has been driven by natural resource development, the oil and gas sector, and the mining sector. The IFC, which establishes lending standards for most large development loans worldwide, the International Council on Mining and Metals, and the International Council on Oil and Gas Producers have issued guidance and standards on HIA (ICMM, 2010). According to Dr. Wernham, industries recognize HIA as a good business practice; HIA provides an opportunity to lower business costs, protect workers and the community, and proactively manage risk. HIAs are often part of an industry's corporate social responsibility plan.

HIAs in the United States

HIAs are occurring across the United States and a few are being developed to address energy and natural resource development decisions (Health Impact Project, 2011):

- biomass (California, Massachusetts, Oregon, Virginia),
- oil and gas leasing (Arkansas),
- mining (Arkansas),
- shale gas development (Colorado), and
- wind energy (Oregon).

According to Dr. Wernham, what is unique about the energy and natural resource sectors is that often these projects are polarizing and political as they focus on issues of jobs, national security, and environmental concerns. There is a mix of environmental health and socioeconomic risks and benefits. Some people in the community are concerned about protecting their communities against environmental risks, whereas other individuals are interested in the economic benefit. Governments also have both an economic interest and an interest in protecting people. Finally, industry has invested years and resources in exploration and planning. Thus, it can be a challenging environment in which to conduct HIAs, he said.

HIA encompasses the breadth of health and it is crucial that HIA include areas other than pollutants. The common subset of influences on health that may be included in HIAs are air quality, water quality, noise, subsistence and agricultural uses, demographic changes and influx of workers, traffic patterns, revenues, and employment and income.

Common questions raised about HIA are: What is the value added of HIA? How does HIA add to what is covered in health risk assessment or an environmental impact statement (EIS)? Dr. Wernham explained that in an EIS or environmental permitting, for example, there are criteria hazardous air pollutants (HAPs) regulated by the EPA; an evaluation is conducted to determine what the specific project will add to the airshed with respect to HAPs (EPA, 2013). HIA provides additional information; it will capture the prevalence of relevant diseases, affected populations, and the populations' relationship to emission sources. Other potential information could be modeling of the emissions to determine which communities may be affected and local concerns from the communities. Understanding local concerns of perceived contamination can be important for understanding changes in behavior that may have health implications. For example, concerns about perceived contamination of air and water in areas reliant on local subsistence farming and hunting may result in individuals' reducing their consumption of wild foods. This change in behavior can have nutrition and health implications.

Dr. Wernham emphasized that HIA is primarily a qualitative assessment, but some analyses can be quantitative. Assessments that look at air and water quality can often be similar to risk assessments, but HIA goes further by finding ways to address local concerns and data gaps. When assessing noise, for example, HIA offers mitigation measures such as sound walls, housing modifications, changes in truck routes, and hours of operations. Traffic is also given a more robust assessment in HIA than in an EIS or health risk assessment. HIA will include the baseline injury rates, identification of the most dangerous intersections or dangerous roads in the community at baseline, and locations of high-risk groups (e.g., school crossings). As part of the process, traffic flow is used to predict potential hotspots. Figure 2-1 shows an increase in traffic injuries in relation to the oil and gas boom in Sublette County, Wyoming. With good planning and management, there are opportunities to prevent these increases in injuries.

Similarly, when assessing demographic change such as the influx of nonresident workers into an area, HIA can provide information on the strain in local services, changes in violence and crime, and changes in the spread of infectious diseases. The IFC has issued guidance documents for managing influx, and the literature supports the need to pay close attention to the management of influx. One of the important contributing factors for the spread of multidrug-resistant tuberculosis in Africa is mining. Partnerships between the mining corporation's occupational health program and the public health infrastructure in the community, for example, could be leveraged to ensure that people have access to directly observed therapy for the treatment of tuberculosis both onsite and in the community (Stuckler et al., 2011).

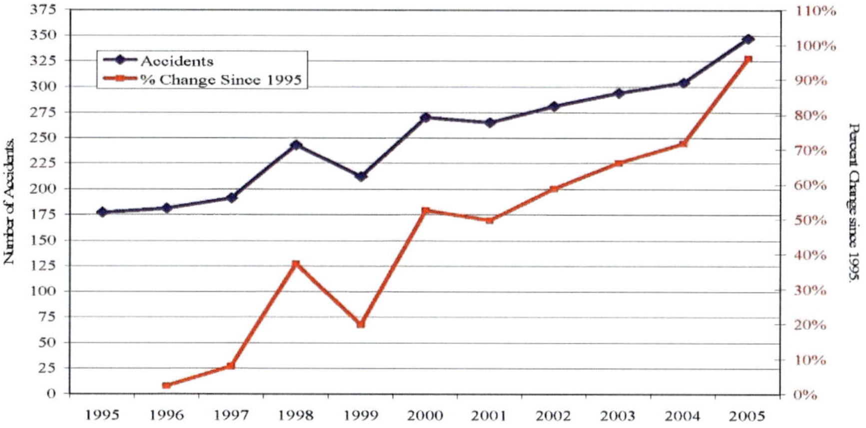

FIGURE 2-1 Increase in vehicle-related accidents in relation to oil and gas boom in Sublette County, Wyoming.
SOURCE: Ecosystem Research Group, 2007. Reprinted with permission from the Ecosystem Research Group.

According to Dr. Wernham, although this type of outcome may not occur in the United States, it raises the point that sometimes the most important health risks are not the ones that are obvious. In North America, there has been one reported example of a transmitted illness in relation to a resource boom. In Fort McMurray, Canada, a significant spike in local syphilis rates was said to be associated with the oil sands boom (Richardson, 2012). The health department did not plan for this health concern. Working with the companies, prevention strategies could have been developed to forestall and manage the problem instead of having to react after the problem had occurred. Further, there are economic revenues and costs that need to be considered when managing risks and maximizing the potential benefits. Cities and states may find opportunities to improve health by leveraging revenues coming into the community to plan for the costs related to increased education, water and sanitation, public safety, clinics and hospitals, and emergency medical services.

The North Slope of Alaska community provides a good example of planning for revenue capture to fund community services in a rural area. The community worked with the oil and gas industry to determine the costs associated with basic water and sanitation needs, staffing schools, police, and emergency medical services (State of Alaska HIA Program, 2011). Planning for revenue capture allowed the community to develop from a village with no running water and primitive waste removal to a community with a water and sanitation system that could run year round, a system that costs millions of dollars. Through efficient early planning with industry, the community was able to plan for the impact of growth associated with natural resource development in the area.

In closing, Dr. Wernham identified a number of issues and challenges associated with HIA of unconventional shale gas operations. He noted that there is a need to engage polarized stakeholders by supporting productive conversation and building common ground among industry; community groups; local, state, and federal governments; and others. He also noted that there are important data gaps, especially in small towns where the amount and type of emissions and discharges and baseline disease prevalence data are often unavailable. HIA is a way to identify these gaps and build a plan to collect the information. Finally, there are often no clear decision points or a comprehensive planning process where HIA would be more helpful. The permitting process is being done one well at a time and not as part of a comprehensive planning process. Dr. Wernham noted that there is not a federal EIS and that many states have not undertaken a comprehensive review prior to permitting.

DISCUSSION

Dr. Portier began the discussion by asking Dr. Wernham for a rough estimate of the resources needed to conduct HIA for shale gas extraction, for example, the time, effort, and human and financial resources required. Dr. Wernham stated that it depended on the scale of the HIA. When designing the HIA, consideration is given to the available resources and time frame available. If the HIA is required in 3 months, an effort is made to give some useful input within this time frame. If a comprehensive HIA is needed, for example, to assist a state developing a new regulatory framework, such as the State of New York, a year is a good estimate. With respect to staff, one person to manage the HIA and a few consultants may be reasonable. HIAs have been conducted for under $100,000, but up to $300,000 may be reasonable to bring together all the right stakeholders.

Carlos Santos-Burgoa from the Pan American Health Organization asked Dr. Wernham to comment on HIA and equity issues and the time horizon used in HIA. Dr. Wernham responded that although he did not emphasize equity in his remarks, HIA certainly does emphasize health disparities or health equity issues. It is emphasized in practice as part of the identification of vulnerable populations. This would include identifying individuals who are vulnerable because of a preexisting health issue, disparity such as low income or ethnic minority status, or earlier exposure to other sources of pollutants. With respect to the time horizon, Dr. Wernham stated that it will depend on the extent that the data allow. Often the time horizon attempts to consider the time span needed for a chronic disease to develop, for example, noncommunicable diseases such as obesity and diabetes or other risks that could take place over 20 to 30 years. The need to consider multigenerational risks and HIAs has been raised.

Dr. Wernham also addressed a number of questions from the audience; one question focused on the possibility of a federally sponsored EIS on hydraulic fracturing. Dr. Wernham responded that what is most important is the value added by the HIA and the deliberative process opportunity it provides. Through the process, critical stakeholders are convened to discuss a proposed action or possible scenarios for action prior to decision making. Another commenter asked Dr. Wernham whether there were true data gaps or if what appeared to be data gaps were in fact a lack of data sharing among the health and environmental communities and industry. Dr. Wernham stated that collaboration among these sectors is valuable; such collaboration is facilitated by having stakeholders involved in the HIA process and dialogue. Through the process, data are shared, data gaps are identified, and data collection activities can be organized. He also noted that in some cases data gaps do exist.

REFERENCES

CDC (Centers for Disease Control and Prevention). 2012. National Biomonitoring Program: About the program. http://www.cdc.gov/biomonitoring/about.html (accessed May 30, 2013).

Ecosystem Research Group. 2007. Sublette County socioeconomic impact study: Draft report. http://www.sublettewyo.com/DocumentCenter/Home/View/363 (accessed May 30, 2013).

EPA (U.S. Environmental Protection Agency). 2013. Air permits: Definitions of selected permitting terms. http://www.epa.gov/region9/air/permit/defn.html (accessed May 30, 2013).

Executive Order 13605. 2012. Supporting safe and responsible development of unconventional domestic natural gas resources. http://www.gpo.gov/fdsys/pkg/FR-2012-04-17/pdf/2012-9473.pdf (accessed May 30, 2013).

Health Impact Project. 2011. Health Impact Project: Advancing smarter policies for healthier communities. http://www.healthimpactproject.org (accessed May 30, 2013).

ICMM (International Council on Mining and Metals). 2010. Good practice guidance on health impact assessment. http://www.icmm.com/library/hia (accessed May 30, 2013).

IOM (Institute of Medicine). 2006. *Rebuilding the unity of health and the environment in rural America*. Washington, DC: The National Academies Press.

NRC (National Research Council). 2011. *Improving health in the United States: The role of health impact assessment*. Washington, DC: The National Academies Press.

Richardson, A. 2012. Oilsands workers at high risk for STI: Report. http://www.fortmcmurraytoday.com/2012/01/30/oilsands-workers-high-risk-for-stis-report (accessed May 30, 2013).

State of Alaska HIA Program, Department of Health and Social Services. 2011. Health impact assessment: Point Thomson Project. http://www.epi.alaska.gov/hia/PointThomsonCompletedHIA.pdf (accessed May 30, 2013).

Stuckler, D., S. Basu, M. McKee, and M. Lurie. 2011. Mining and risk of tuberculosis in sub-Saharan Africa. *American Journal of Public Health* 101(3):524–530.

White House. 2012. Investing in America: Building an economy that lasts. http://www.whitehouse.gov/sites/default/files/investing_in_america_report_final.pdf (accessed May 30, 2013).

3

Geographic Footprint of Shale Gas Extraction

This chapter provides a summary of presentations on the process of shale gas extraction and its potential geographic impact. The first presentation provides an overview of the characteristics of a shale gas production site, the method of hydraulic fracturing, and the safety procedures used by one company. The second presentation describes federal efforts to assess the effects of shale gas production on U.S. geography, or its "geographic footprint." Both presentations feature examples using the Marcellus Shale, a source of natural gas reserves located in the Appalachian Basin. The presentations are followed by a summary of the discussion that ensued.[1]

FRACTURING: ACCESSING SHALE AND TIGHT GAS

David Cole, M.S.
Regional Discipline Leader—Production Technology/Chemistry
Shell Upstream Americas

David Cole began by noting that hydraulic fracturing is, at its most basic, a process of pumping fluid into a rock faster than that rock can absorb the fluid. This results in cracks that can be held open by the injection of a solid material in order to extract the gas and oil resources in the rock. Fractures such as these can occur naturally, although some rock layers, such as tight shale layers, are naturally impermeable to fluids and gas. Since the 1940s, hydraulic fracturing has been used to extract oil or gas from the tight shale layers of rock, and this practice has been used in more than a million wells in the United States alone. Primarily, this

[1] Dr. Charles G. Groat gave a presentation on "Assessing the Perceived and Real Environmental Consequences of Shale Gas Development: Report on an Initiative of the Energy Institute, The University of Texas at Austin." A summary of that presentation is not included here because of questions that have arisen regarding the conduct of that study.

process is used for extracting shale gas, but there has been movement toward extraction from oil-bearing shales as market demand increases.

Hydraulic fracturing relies on the sophisticated use of pumps to create a pathway into the rock. Sand or an engineered ceramic material is then placed into the pathway to keep the cracks open after the hydraulic pressure is removed. Mr. Cole stated that the typical dimension of a hydraulic fracture is microns to 0.25-inches wide, 500- to 2,000-feet long, and 20- to 400-feet high, depending on the geography.

Thirty years ago, there was typically 1 well for every 40 acres, which translated to approximately 16 well locations in a square mile, connected by gravel roads. With advances in technologies, there is often now one location per square mile and all the wells are drilled in the direction of the least principal horizontal stress. For example, wells near mountainous regions will run parallel to the mountain. Newer technologies allow engineers to use a steel drill pipe to turn and bend in order to orient the well in any direction. Mr. Cole noted that clustering wells to one surface location has a number of advantages, including reducing the number of trees cut down, reducing traffic, and reducing emissions. Nonetheless, each drilling site will affect the surface geography with wells, roads, and supporting facilities.

Mr. Cole explained that careful well planning is crucial to isolate the fluids in the well and avoid contamination of drinking water. Different companies have varying strategies for water management; for example, Shell captures the water used in the fracturing process and places it into tanks with secondary containment for recycling. Other companies may use lined pits to capture this fluid, and the location is chosen based on knowledge of the depth of the groundwater and other local receptors. When planning a new well location, Shell measures the resistance to an electrical current in order to determine where fresh water is located, so that efforts can be taken to protect this water when building the well.

The first step in drilling is to put in a conductor pipe, which is a structure designed to carry the load, akin to the foundation of a house (see Figure 3-1). Additional steel casing strings, blowout preventers, and other equipment are installed through the surface using a drilling mechanism and are cemented into place. The surface casing string, which consists of steel pipes coupled together with screws, is lowered into a drilled hole that runs the depth of the freshwater layer, to protect the groundwater. Cement is pumped down into this casing to seal it into place. Check valves on the bottom of the casing will help to prevent contamination and flowback and preserve isolation of the groundwater. The casing and the cement that make up the wellbore are tested to meet strict specifications of integrity before the drilling begins. Intermediate casing may be necessary, depending on where the drilling takes place.

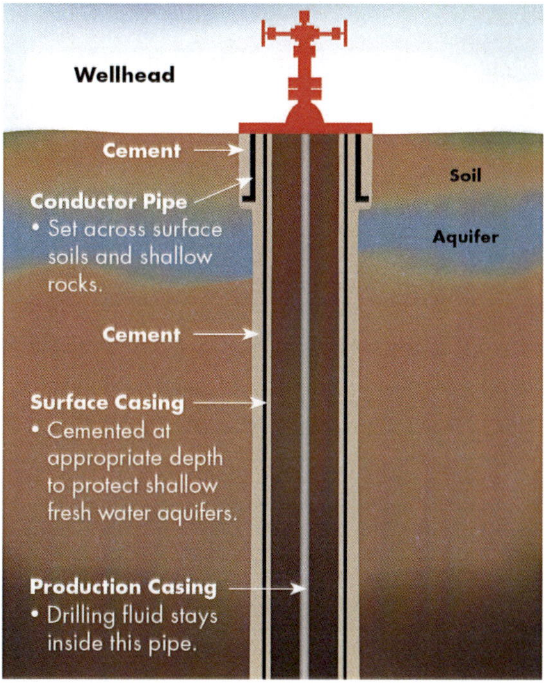

FIGURE 3-1 Illustration of wellhead and casings.
SOURCE: Shell, 2013. Reprinted with permission from Royal Dutch Shell.

After installation of the surface casing and any intermediate casing, the production interval is drilled. This smaller steel pipe is used to transport the fracturing and produced fluids. The hole for the production casing is drilled in a vertical direction until the correct depth is reached; then, the drilling direction will turn horizontally for thousands of feet before the target formation is reached. This type of long, horizontally drilled well is common in the Marcellus Shale, which composes much of the Appalachian Basin region. The casing string is inserted and, similar to the surface casing, cement is pumped into the casing to isolate it. Finally, the well is pressure tested to ensure its integrity.

To access the formation, mechanical punches produce one-quarter- to three-eighths-inch holes in the steel production casing. It is through these holes that pressurized jet streams of drilling fluid will create the fractures in the rock formation. Throughout the life of the well, pressure sensors are used to check for a firm seal. The well is then prepped for fracturing by cleaning out the casing with water. During the first release of fluid, considerable water and solids are produced. A temporary production facility is installed to separate the water from the solid waste. (The composition of the fracturing fluid is discussed in more detail in Chapter 5.)

Mr. Cole stated that fracturing is an engineered process that takes into account the strengths and properties of the target rock formations. Understanding these properties makes the fracturing of rocks predictable and consistent. Microseismic listening techniques are used as fracturing fluids are pumped into a well. The sound of rock breaking is an indication of where activity is happening underground. This technique is used to optimize gas development, because it allows the engineers to know the orientation and length of the fractures. Further, as more wells are drilled, this information helps to plan for future well placement. One of the side benefits is that microseismic events allow for a company to have an indication of dimension. After the gas or oil is extracted, it is transported via pipeline.

The primary concerns in the development of these wells are ensuring that groundwater is protected and that gas and oil are not lost during the process, said Mr. Cole. Fracturing activity takes place thousands of feet from freshwater, and securing the casing with cements helps maintain isolation from drinking water sources. Transparency of information about hydraulic fracturing is also an important consideration. Websites such as Fracfocus.org, a chemical disclosure registry operated by the Groundwater Protection Council, provide a voluntary reporting site for each well (GWPC and IOGCC, 2013). Mr. Cole stated that Shell has reported every well since January 2011 and sees these efforts as best practices for the industry. Additionally, Mr. Cole reported that Shell operates under a series of principles to

- ensure the safety of workers and well integrity,
- conduct operations to protect groundwater and reduce water use as reasonably practicable,
- protect air quality and control fugitive emissions,
- work to reduce the operational footprint, and
- engage with local communities regarding socioeconomic impacts that may arise from Shell operations.

Discussion

Following Mr. Cole's presentation, Roundtable and audience members were invited to ask questions. Christopher Portier began by asking if Shell has ever done a health impact assessment for shale gas extraction, to which Mr. Cole replied "yes." Bernard Goldstein asked if the tremendous increase in the ability to extract natural gas would continue, and if states that began these activities earlier are receiving as much economic benefit as those that waited for improved extraction technologies. Mr. Cole replied that it has always been known that these resources existed, and only recently was it learned that it could be turned into reserves and, in turn, produced. Because development is ongoing and companies are continually evaluating and looking for more efficient ways to extract shale gas, there is likely no problem for states that started early. Richard Jackson asked for more details on what the wells are producing (i.e., oil) and how shale gas is extracted and removed from the sites. Cole described the process, which began with natural gas in the Barnett Shale and a movement toward oil-bearing shales, such as in North Dakota. The movement toward oil extraction is driven by price differentials on an energy basis between oil and gas. Natural gas and oil are both pipelined from the sites, though smaller developments can truck the resources out.

An audience member questioned if there was a consensus in the industry on what level of transparency they are willing to provide to the public on fracturing fluids. Mr. Cole reiterated that Fracfocus.org is a valuable source for such information, and has more than 70 different companies with 9,000 wells reporting to that database. Mr. Cole also mentioned that the service companies' trade secrets may not allow complete reporting, but some states have required that this be reported to them in case they need to respond to a related problem. Most of the fracturing fluid chemicals are not secret, but the formulations of these chemicals are proprietary. Linda McCauley asked about the range of the size of the industry, particularly the number of smaller companies. Mr. Cole noted that major oil companies make up a small fraction of the overall oil produced in the United States, and that the industry is dominated by large independents. An audience member via webcast asked how many Marcellus Shale wells have been drilled in multiple directions, to which Mr. Cole replied that almost everything in the Marcellus Shale is horizontally drilled. Another audience member inquired further on the fluids used in fracturing and expressed concern about the composition of the fluids (e.g., if the chemicals are endocrine disruptors) and the protection of water. Mr. Cole restated that he is not an expert on the chemicals used in these fluids and could not speak to that, and that the real issue for safety is in isolating the fluids being pumped in and out and maintaining the well's integrity. From Mr. Cole's perspective, these issues of concern are not happening. In response to another question about how the target shale layer is drilled, Cole responded that directional drilling techniques allow the engineers to always know where the well is relative to the target location. In some cases, geosteering tools can be used.

The next question from the audience addressed drilling intensity rates. The audience member recalled learning that wells for fracturing tend to produce a lot of gas in the first year, and then production declines, and that refracturing a well does not produce as much the second time around. In addition, it is suggested that there is a time limit as to how quickly a company needs to drill once it is granted a lease, creating a push to drill faster. Mr. Cole defined the drop-off in production as "hyperbolic decline," which occurs as new wells are drilled and experience a peak, followed by a decades-long tail of production. For the final question, an audience member asked if Shell conducts testing on water sources before fracturing, to which Mr. Cole replied "yes." Shell tests water sources before drilling and is committed to proactive testing after a certain amount of follow-up time.

GEOGRAPHIC FOOTPRINT

Michael Focazio, Ph.D.
Assistant Program Coordinator,
Toxic Substances Hydrology Program
U.S. Geological Survey

Michael Focazio began by noting that human activity generally affects the environment in some way, whether it be as benign as a nature hike or as substantial as clearing a field to plant corn. The evidence of those activities is referred to as a geographic footprint. Environmental health faces a challenge in measuring the impact associated with these footprints. The U.S. Geological Survey (USGS) has endeavored to perform scientific assessments of geographic footprints, both spatially and temporally.

Dr. Focazio explained that one way the USGS measures geographic footprints is by surveying the aerial extent of the land and observing land changes associated with the activities. This method has been used to describe the impact of shale gas extraction on land surrounding a well site. As detailed previously in Mr. Cole's presentation, the well site for hydraulic fracturing involves roads, trucks, water storage, and surface drilling, which contribute to the geographic footprint of that activity. U.S. shale gas extraction increased more than fourfold between 2007 and 2011 (GAO, 2012). As the activity increases, the geologic extent of these recoverable resources, or sources of shale gas, becomes included in the definition of the extraction's geographic footprint and expands beyond measurements of surface land change.

Dr. Focazio stated that the geographic footprint of shale gas extractions can be conceptualized on three levels: national, regional, and local. Nationally, an assessment measures the extent of technically recoverable resources: where they occur and how frequently. It is useful to compare it with the superimposed map of North America that shows the geologic extent of the resource, as displayed in Figure 3-2.

On a regional level, the geographic footprint factors in roads, pipelines, and other infrastructure developed to extract and transport the gas. Pipelines in particular are a major component of the infrastructure

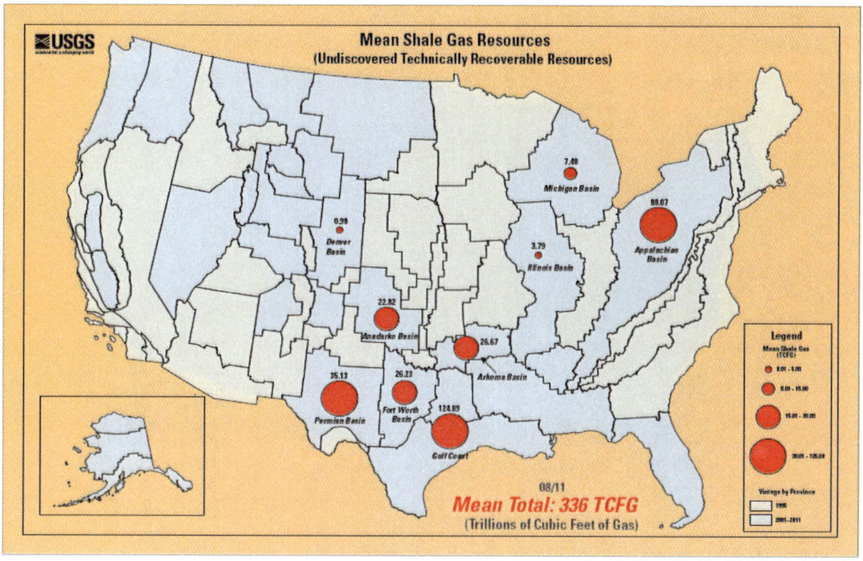

FIGURE 3-2 U.S. mean shale gas resources.
SOURCE: USGS, 2012.

with potentially tremendous impact when crossing through sensitive ecosystems. The local impact includes surface-level activity, such as land clearing and the construction of well pads or water storage facilities, which entails well site operations, ponds, and roads at the extraction site. The design of a shale gas extraction site is similar to what it is being used in oil and natural gas extraction, in terms of vehicles for transportation, the derrick, water lagoons, and water sources.

The USGS used an example of activity in the Marcellus Shale to describe a geographic footprint, said Dr. Focazio. Mostly permits for oil and gas have been issued although coal mining and methane extraction are also occurring in this region. The map in Figure 3-3 shows the extent of the number of permits that have been issued. Technically, this is equivalent to the number of sites expected, and they are located all along the Marcellus Shale, ranging across New York, Ohio, Pennsylvania, Virginia, and West Virginia. Most of the permits are not being used, but the potential geographic footprint can be assessed prospectively.

The aerial view in Figure 3-4 is the before (2006) and after (2010) development of a well pad site. From this image, it is clear that the land has been cleared, and that roads and well pads have been added. In itself, the clearing of the land may have important ecological impacts. Dr. Focazio noted that in the Marcellus Shale, the area that is covered by these sites averages 7 acres, ranging from 5 to 10 acres total for shale gas

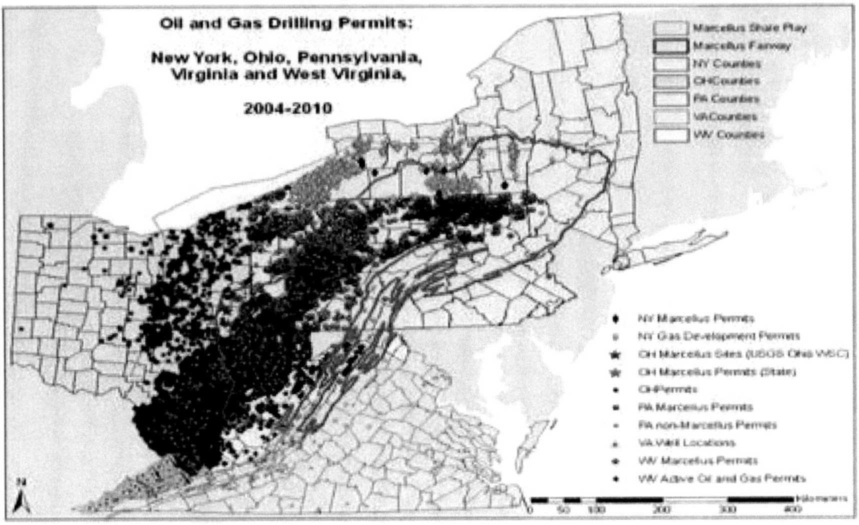

FIGURE 3-3 Drilling permit locations.
SOURCE: USGS, 2012.

sites. In Pennsylvania, there are approximately 10,000 permits issued for shale gas extraction. The approximate land area of that geographic footprint is 76,000 acres. This is a very small percentage (0.3) of the land area in Pennsylvania, and not all of this is currently being drilled.

After considering the local impact on the land, Dr. Focazio noted that it is important to understand the factors that can expand the activity to a regional level. A network of roads needs to be developed in order to load, unload, and transport the shale gas. Vehicles moving at an increased frequency than before and traveling to locations that had not been accessed before construction of the well pad may have a tremendous impact on the geography. The well pad will also require pipelines and the infrastructure to move fluids. Processing sites will also be created and, in most cases, those are centralized facilities that differ from the well pad and are often larger.

Most of the attention around shale gas extraction has focused on understanding the human health consequences, said Dr. Focazio. Assessments of the geographic footprint are primarily concerned with the measurable ecological impacts of the activities. In many cases, these concerns will overlap. The clearing of land will affect native species, the increase in transportation will change the air composition, and the increased use of water will affect hydrological cycles. This can be taken into perspective if we consider that most concerns from the activity have been driven by the proximity to residential areas.

FIGURE 3-4 Local footprints of Marcellus Shale gas well sites from 2006 (left) and 2010 (right).
SOURCES: Slonecker et al., 2013.

The way that geographers have been monitoring the issue is by utilizing programming computers to detect land changes in a certain area. The computer is configured to recognizing patterns; for example, identifying how much fragmentation appears in the land or more sensitive details such as mapping well pads. An important aspect of this kind of monitoring methodology is time. Acreage compares the differences before and after the well pad has been constructed, but pattern recognition programs have the capability to record changes over time. After the well has been constructed, the program, for example, can identify the density of the forest before and after a shale gas production site is constructed, which may entail reducing forest to build pipelines.

Offsite activities and infrastructure also contribute significantly to the geographic footprint of hydraulic fracturing. Use of sand to fill the fractures is a characteristic of shale gas production that makes sand mining an important consideration. It is an operation that requires an extensive infrastructure and that covers a large expanse of land. For the Marcellus Shale sites, sand mining is mostly done in the Midwest, primarily in Minnesota and Wisconsin, and transportation is mostly done by train. Sand mining has grown exponentially in the last decade with the increase in demand of natural gas and, as a consequence, the geographic footprint of shale gas extraction includes an even larger region.

At this time, there is no objective conclusion about what constitutes an unacceptable geographic footprint for shale gas extraction, said Dr. Focazio. Over time, geographic scientists will strive to gain a better understanding of the long-term impacts. Current shale gas resources, such as the Marcellus Shale, are finite. If demand remains high, production may focus on deeper sources such as the Utica Shale, which is located underneath the Marcellus Shale and not currently used in Pennsylvania.

Discussion

Dr. Focazio emphasized the importance of a balanced, scientific perspective when evaluating the impact of shale gas extraction in response to Roundtable members' questions about the promotion of misinformation and the ubiquity of natural gas wells in certain locations. One audience member asked about the extent of radioactive releases during hydraulic fracturing. Dr. Focazio responded that the geology of the shale structure dictates some of the effects of the extraction. For example, radionuclides naturally occur in the rock formation and may be released in the course of fracturing. It is unclear if hydraulic fracturing mobilizes more radioactive releases than would normally be produced; however, analyses are under way to examine the water produced in the extraction process for the presence of radionuclides and other chemicals. Another audience question centered on asbestos-type molecules and radiation, and asked if there has been any measurement of those at hydraulic fracturing sites. Dr. Focazio replied that these data are collected to characterize aquifer water quality generally, including other chemicals, but are not collected specifically for geographic footprinting. Throughout the discussion, Dr. Focazio reiterated that measuring the geographic footprint of shale gas extraction requires science that considers the comprehensive activities of hydraulic fracturing and the related health and safety consequences.

REFERENCES

GAO (U.S. Government Accountability Office). 2012. Oil and gas: Information on shale resources, development, and environmental and public health risks. Washington, DC: GAO.

GWPC and IOGCC (Ground Water Protection Council and Interstate Oil and Gas Compact Commission). 2013. FracFocus Chemical Disclosure Registry. http://www.fracfocus.org (accessed May 30, 2013).

Shell. 2013. Hydraulic fracturing: Your questions answered. http://s00.static-shell.com/content/dam/shell/static/usa/downloads/onshore/abc002-hyd-frac-insert0623.pdf (accessed May 30, 2013).

Slonecker, E. T., L. E. Milheim, C. M. Roig-Silva, and A. R. Malizia. 2013. Landscape consequences of natural gas extraction in Allegheny and Susquehanna counties, Pennsylvania, 2004–2010. Open-file report 2013-1025. Reston, VA: U.S. Geological Survey.

USGS (U.S. Geological Survey). 2012. National oil and gas assessment 2013 assessment updates. http://energy.usgs.gov/OilGas/AssessmentsData/NationalOilGasAssessment/AssessmentUpdates.aspx (accessed May 30, 2013).

4

Occupational Health and Community Impacts

This chapter provides a summary of presentations focused on the communities impacted by shale gas extraction, concerning the occupational opportunities and hazards for populations and the everyday consequences of living near a hydraulic fracturing site. The presentations describe separate formal assessments conducted at or around shale gas production sites in Colorado and other states, while the final presentation focuses on an overview of the economic impacts in Pennsylvania. The presentations are followed by a summary of the discussion that took place.

NIOSH FIELD EFFORT TO ASSESS CHEMICAL EXPOSURES IN OIL AND GAS WORKERS: HEALTH HAZARDS IN HYDRAULIC FRACTURING

Eric J. Esswein, M.S.P.H.
Senior Industrial Hygienist,
National Institute for Occupational Safety and Health
Centers for Disease Control and Prevention

Eric J. Esswein began his presentation by noting that occupational safety hazards in oil and gas extraction (including hydraulic fracturing) are known but the same cannot be said for occupational health hazards, especially risks for chemical exposures. Attention has typically focused more on safety—work practices, policy and procedures to prevent acute injury—than on health (understanding risks for chemical exposures and prevention of chronic disease). The National Institute for Occupational Safety and Health (NIOSH) implemented the *NIOSH Field Effort to Assess Risks for Chemical Exposures in Oil and Gas Workers*[1] to better understand risks for chemical exposures in workers involved in oil and gas extraction activities.

---

[1] See http://www.cdc.gov/niosh/docs/2010-130 (accessed May 30, 2013).

Operators within the oil and gas industry have been voluntarily participating with NIOSH since 2008 and since 2010 in the NIOSH *Field Effort*. Participation occurs through memoranda of understanding between NIOSH and various companies. NIOSH researchers' review work practices and chemical products used on site, and conduct industrial hygiene exposure assessments to determine if occupational health risks might be present. Depending on the outcome of the field research studies NIOSH provides recommendations to control exposures if risks are determined.

Mr. Esswein provided a list of some chemicals that may pose exposure risks during oil and gas extraction including respirable crystalline silica, diesel particulate, volatile organic compounds (VOCs), hydrogen sulfide, acid gases, aldehydes, and elements (i.e., lead) from lead-based pipe thread grease. Although the NIOSH field work in 2010 and 2011 excluded physical agents (i.e., noise, radiation, etc.), the research team looked at chemical exposures across a wide range of oil and gas extraction basins in the United States. NIOSH visited 11 hydraulic fracturing sites in five states, during different seasons and at varying elevations. Respirable crystalline silica (from quartz sand) was identified as a potential occupational exposure hazard at hydraulic fracturing sites. Sand is used as a proppant to hold open fissures and cracks in the hydrocarbon-bearing formation that are created by hydraulic fracturing.

NIOSH determined that respirable crystalline silica presented an occupational exposure hazard likely greater than exposures to chemicals used during hydraulic fracturing. Crystalline silica is the known causative agent for the lung disease silicosis and is associated with lung cancer. Silicosis is preventable by eliminating exposure, but discontinuing exposure does not stop progression of the disease, which is incurable, irreversible, and progressive. Industry wide, in the United States, exposures to crystalline silica are associated with 160–200 worker deaths annually (Rosenman et al., 2003); consequently occupational exposures are regulated by the Occupational Safety and Health Administration (OSHA). Millions of pounds of quartz-containing sand are used during hydraulic fracturing. Sand-moving transport operations result in respirable crystalline becoming airborne during handling and pneumatic transport by a variety of machines used during the hydraulic fracturing process (e.g., sand movers, sand transport belts, and sand and water blender trucks). Windy conditions on-site can also contribute to dust generation and exposures to unprotected workers.

During 2010–2011 the NIOSH *Field Effort* research team collected 111 full-shift personal breathing zone samples for respirable crystalline silica during active hydraulic fracturing operations. NIOSH determined that more than 50 percent of the samples exceeded a calculated OSHA permissible exposure limit, 68 percent were greater than the NIOSH recommended exposure limit; in some cases by factors of 10 or more, which exceed the assigned protection factor for the typical type of respirator (half-face elastomeric or filtering face-piece) used by the workers at the sites visited. As part of the study, NIOSH identified seven primary points of dust generation and developed control recommendations that included both passive and active controls. The NIOSH team also described nine possible interventions to control dust, including

- consideration of Prevention through Design[2] on future versions of sand moving equipment that has built in dust control;
- use of remote operations to keep workers out of areas of high dust concentrations;
- substitution of sand with ceramic or other proppant materials;
- installation of active controls such as the NIOSH-designed mini-baghouse retrofit assembly (technology to reduce amount of dust released), over the thief hatches (access ports over sand movers), and consideration of enclosed sand transport mechanism such as a screw augur retrofit assembly to replace belt transport of sand;
- use of enclosures around bottoms of sand movers (such as stilling or staging curtains) and the dragon tail (end of the sand belt) on sand movers;
- minimizing the distance that sand falls from the dragon tail;
- use of end caps on fill nozzles on sand movers;
- use of amended water for dust control on site; and
- implementing an effective respiratory protection program.

Mr. Esswein reiterated that NIOSH determined that respirable crystalline silica is a significant occupational health hazard associated with hydraulic fracturing; diesel particulate is also a likely occupational health hazard.

Discussion

Following the presentation from Mr. Esswein, Roundtable members started the discussion by asking about silica exposures at mines during the loading and unloading of trucks. Mr. Esswein noted that NIOSH's assessment was limited to evaluating exposures only at oil and gas sites, not any further upstream. Nsedu Witherspoon inquired about exposures to pregnant workers and Mr. Esswein replied that he only saw two female oil and gas workers, and the NIOSH team did not evaluate their exposures. Bernard Goldstein asked about a 2008 article showing an increase in injuries per shale gas well and asked about the status of injuries. Dr. Goldstein went on to comment on integrating workers' health and

---

[2] Prevention through Design is a NIOSH-led effort to address occupational safety and health needs in the design process to prevent or minimize the work-related hazards and risks (http://www.cdc.gov/niosh/topics/ptd [accessed May 30, 2013]).

environmental health, and asked whether there is any indication that companies that are less careful with the environment are also less careful about worker health, that is, determining whether worker health violations track with the same companies that are fined for environmental risks. Mr. Esswein replied that oil and gas extraction and production have about seven times the fatality rate of general industry (27–28 per 100,000 compared with 4 per 100,000), and that most injuries are attributable to motor vehicle accidents, although there are additional deaths from causes such as being struck by equipment, being caught between parts of equipment, and falling from heights. Mr. Esswein also acknowledged the challenges represented by a worker population that is highly transient, which makes surveillance and follow-up efforts difficult.

Mr. Esswein responded to a question about dissemination of findings by noting that NIOSH participates in the oil and gas workgroup of the American Industrial Hygiene Association, and that the NIOSH team will participate in the OSHA Safety Conference, which will focus on oil and gas. They have submitted an article for the *Journal of Occupational and Environmental Health* (Esswein et al., 2013). An audience member inquired whether NIOSH researchers monitored the exposures of female security officers stationed outside hydraulic fracturing project sites. Mr. Esswein stated that the security staff he has seen are typically stationed on the lease road some distance away from the wellhead, and NIOSH had not researched these exposures. Another person asked whether NIOSH's visits are announced; Mr. Esswein replied that NIOSH is a research agency, not a regulatory agency, and site visits are scheduled, not unannounced. Moreover, it seems implausible that operators at a hydraulic fracturing site could clean up before a NIOSH visit because they have a performance contract to fulfill, and they need to pressure up and pump, making it difficult if not impossible to somehow stop and spruce things up.

One audience member inquired whether the NIOSH researchers have reviewed data to show whether hydraulic fracturing site workers are experiencing silica-related diseases, and secondly, whether they have looked at other types of silica-related illnesses, including autoimmune, chronic respiratory, or kidney diseases. Mr. Esswein stated that NIOSH has not done any biological monitoring of workers, and reiterated that the work was an exposure assessment study, not a health study, and the study was still progressing. There is no systematic surveillance for oil and gas extraction workers and no known sentinel cases of accelerated or acute silicosis in the workers. Companies have informed NIOSH that they are expanding preemployment physicals to assess baseline pulmonary function and improving respiratory protection programs. Some companies have begun considering or even implementing some of the NIOSH recommended controls to limit exposures to crystalline silica dust. Another audience member asked if silica dust is going offsite or if this was an issue, and if NIOSH had collected air samples offsite; Mr. Esswein responded that sampling was limited to workers in the immediate area around the well pad. Mr. Esswein noted that he considered the sampling conducted to be fairly representative of the workforce at hydraulic fracturing sites (from the "company man" to the variety of contractors on site), because most workers were willing to be sampled. Another audience member asked whether any workers were represented by unions, and Mr. Esswein answered that they were not.

COMMUNITY IMPACTS OF NATURAL GAS DEVELOPMENT AND HUMAN HEALTH

Roxana Witter, M.D., M.S.P.H., M.S.
Assistant Research Professor, Environmental and Occupational Health
Colorado School of Public Health

Roxana Witter noted that she led a team from the Colorado School of Public Health that conducted a health impact assessment (HIA) of a proposed natural gas project in a small community in western Colorado. The research team identified three possible types of exposures: chemical exposures, nonchemical exposures such as industrial activities (e.g., noise, traffic), and community changes. The team made more than 70 recommendations to reduce the risk of effects, including noise, traffic, and community changes (Witter et al., 2011).

Dr. Witter explained that some of the data reviewed in the HIA were provided by operators who collected noise data, such as the noise emitted by diesel generators with and without noise blankets. Although it is unclear what the cumulative noise of 20 wells being drilled might be, and the effects of sustained noise over a period of time, noise levels can disturb sleep, affecting cognition, mood, and school performance.

In thinking about the potential effects of truck traffic, the researchers recognized that it meant not only traffic around a well pad and its neighboring homes, but also a broader range of effects dispersed around the community, including exposure to exhaust, vibration, and dust, and safety risks, especially along haul routes that might be traveled by children.

In assessing the community changes, the research team examined the social disruption, with the potential to affect quality of life, elicit stress and anxiety, and pose other risks to physical and mental health, as well as safety. The data reviewed in the HIA were from nearly a decade of natural gas projects in nearby areas of western Colorado. Between 2005 and 2008, well drilling was scaled up rapidly, but a steep decline took place in 2009, perhaps due to the drop in natural gas prices. The researchers reviewed rates of police arrests and sexually transmitted diseases, finding patterns that coincided with the increase in the introduction, expansion, and then decrease in drilling activity.

Dr. Witter noted that although there are some gaps in knowledge, and making the link between natural gas drilling and social effects is not entirely straightforward, there is no doubt that there are some lessons in the available evidence, and that the social sciences and social epidemiology are useful in examining these relationships. For example, the sociology literature on boomtowns in the 1970s' energy development is instructive. In 2009, sociologist Jeffrey Jacquet reviewed this literature and documented a similar "boomtown model" in Wyoming. People living in natural gas areas of Colorado, North Dakota, Pennsylvania, and Wyoming will likely find aspects of the model familiar (Jacquet, 2009). As the energy industry rapidly expanded, such communities saw rapid population influx, unprepared local governments, burgeoning resentment between old and new residents, a need for tax increases to support new and greater infrastructure, boom and bust cycles, inflation, mixed economic effects, and industry monopolization of information, leading to distrust of industry and a sense that there was a power grab away from the community. Other sociology and social epidemiology literature beginning in the 1920s and 1930s document the effects on health of the social circumstances. For example, highly mobile, socially isolated groups (e.g., the families that follow industrial workers to new sites of work) experience high rates of disease associated with social factors such as lack of social cohesion, social capital, trust, and shared values. Dr. Witter stated that evidence indicates that the ties that facilitate collective action confer benefits on health. Multilevel statistical analyses can be conducted to measure the impact of environment on health, and to show the effects of psychosocial stress on the body and community vulnerability contributions to health problems.

The research team explored several potential solutions, for example, determining what boomtown characteristics may be present around natural gas projects, and what characteristics could be amenable to interventions. Are there ways to shape population influx, perhaps through controlled or slower development, involving local governments, facilitating community engagement, and providing jurisdictional control?

The Colorado School of Public Health research team recognized that the environment around natural gas drilling sites is changing and difficult, and that it is hard for HIA to assess all impacts up front, in advance of actual implementation. Establishing monitoring mechanisms and putting in place adaptive management plans can be incorporated into the recommendations of HIA and are important for addressing unexpected and ongoing impacts.

ECONOMIC AND COMMUNITY IMPACTS OF GAS SHALE IN PENNSYLVANIA

Timothy Kelsey, Ph.D.
Professor of Agricultural Economics, State Program Leader
Economic & Community Development
The Pennsylvania State University

Timothy Kelsey noted that the Marcellus Shale is the second largest natural gas field in the world according the U.S. Geological Survey, covering hundreds of thousands of acres. It has considerable economic effects, including local economic benefits, which have led to a great level of interest. Natural gas is a nonrenewable natural resource, and any discussion about jobs and economic development related to natural gas needs to acknowledge that when the gas is gone, the basis for economic activity is gone as well.

There is much uncertainty about health effects and other aspects of unconventional gas development, stated Dr. Kelsey. In a context of great public interest, both positive and negative, there is a quest for scientific answers, and when they are unavailable, people are forced to rely on anecdotal information. This is a difficult basis on which to hold cogent policy discussions.

Dr. Kelsey suggested that gas activity should be thought about holistically, across all three phases: development, production, and reclamation. The development phase is short-lived and labor-intensive, the production phase is long-lived with small and steady labor force requirements, and in the reclamation phase, the employment needs decline. There are two estimates of the duration of drilling in Pennsylvania, either 30 years or 50 years of active drilling, but the number of workers will depend on the number of drilling rigs and number of wells being dug.

Different jobs are needed at each phase, stated Dr. Kelsey, and when the development phase begins, large numbers of jobs will be created, but when the drilling ends, many of the jobs will go away.

Gas development is more than just wells; it involves multiple well pads, supporting many locations, multiple specialized companies and workers, changing transportation patterns, and a major need for coordination and communication. In addition to "short-run" facilities, specifically well pads, there is a need for "long-run" facilities. The former include staging areas; worker housing; office areas; storage, maintenance, and compressor stations; and water withdrawal facilities and treatment areas.

Pennsylvania's economic experience through March 2012 includes hiring, lower unemployment, and higher average wages in Marcellus counties. The state government has seen some increased revenue (taxes), but there are no estimates yet of the total costs to them.

Dr. Kelsey described in some detail the business activity surrounding shale gas extraction in Pennsylvania. Drilling each well requires about 420 individuals across 150 different occupations (Kelsey et al., 2011). For the first well on a pad, there are 13.1–13.3 full-time direct jobs, and 9.65–9.85 full-time direct jobs for each subsequent well (MSETC, 2011). Once the wells are drilled, every 100 dry gas wells generate 19 long-term full-time jobs, and every 100 high-BTU ("wet") gas wells generate 30 long-term full-time jobs (MSETC, 2011). The natural gas workforce is occupationally diverse, ranging from general labor to drivers and from engineers to geologists. About 75 percent of the jobs do not require a college education, although many require some type of certification (MSETC, 2011).

Dr. Kelsey stated that in the decade since drilling began in Pennsylvania, many questions about community and family effects have arisen, but there are many questions yet to be asked, and many answers yet to be found. For example, there is a great deal of discussion about infrastructure needs and how these should be met and about the effects of the shale gas boom on housing affordability and on youth migration. Rents have tripled and even quadrupled in some of the affected counties. This is creating great burdens on vulnerable populations. Schools are reporting a decrease in the number of subsidized lunches, not due to increased income levels but due to families moving out of an area they can no longer afford.

Local infrastructure issues include damage to roads, highway safety, housing costs and availability, sewerage and water requirements, and need for police, emergency services, and schools. On a positive note, stated Dr. Kelsey, development (such as hotels) has also been noted in the Marcellus counties. The shale gas drilling has dramatically increased traffic, with studies indicating that in 2010, some drilling locations averaged 3,100 to 3,900 tractor trailers at each location, a 10-fold increase over several years of drilling. Tractor trailer traffic can have considerable effects on local communities, including implications for road safety.

From a sociologic perspective, researchers are observing more conflict in communities, including polarization, social service demands, early leasers versus late leasers, and newcomers versus oldtimers. From an economic perspective, there are effects on taxes and politics.

Dr. Kelsey emphasized the importance of local engagement, and stated that there is little opportunity for local governments to weigh in on what is happening in their communities. The state recently enacted a law to address this.

In his concluding remarks, Dr. Kelsey emphasized the importance of looking beyond the boom and examining what will be needed after the boom.

DISCUSSION

Following the presentations by Dr. Witter and Dr. Kelsey, Roundtable and audience members were invited to ask questions and engage in discussion. Richard Fenske began by asking Dr. Kelsey why remediation was not on the job chart and how this is handled in the permitting process in Pennsylvania. Dr. Kelsey replied that they are not at that stage yet.

Patricia Verduin asked whether anything can be learned from the boom–bust experiences of other industries. Dr. Witter said that potential lessons could be gleaned from the experience of the energy industry in the West in the 1970s and 1980s. She acknowledged, however, that there are geographic and other differences that should be noted. For example, areas that are less populous have had to bring in a labor force, and in areas with limited infrastructure, workers have had to live next to the well pads. Also, when production sites are dispersed, as in mining, there are lessons that could be learned about what has and has not worked and how to avoid mistakes that were made.

Lynn Goldman restated the assertion made by several speakers that there are gaps in research on the health and social effects of shale gas extraction, and asked about the level of funding available to support research in the field. Dr. Kelsey stated that he has not lacked funding and that much of his work has been supported by the Pennsylvania state government. There are issues related to actual or perceived conflict of interest if industry or an advocacy organization provides the support, and that can taint how the research is received no matter how well conducted or objective. Dr. Kelsey stated his belief that there is not a stable source of funding for long-term monitoring, and the importance of this type of research to better understanding the effects of shale gas extraction. Dr. Witter agreed, and stated that the Colorado HIA relied on available data sources because that was the best option for the funding available. The field is so new that there is no track record that could serve as the basis for funding searches.

Al McGartland remarked that the comment about school turnover made him wonder why companies appeared not to have hired locally and thus helped to maintain community stability. He asked whether they are attracting outsiders because they are paying more and because the skills are not available locally. Dr. Witter stated that some skills are necessary for the work, and she is aware that some companies have been unable to find the skills locally, but she is also aware that in some places in Colorado and Wyoming some companies offered training to local people. Dr. Kelsey found similar company training in Pennsylvania as part of an effort to hire locally, but he also learned that some local businesses were hurt by having employees hired away by the better-paying gas companies. To a follow-up question about why hotels needed to be built, Dr. Kelsey responded that many of the Marcellus counties have old housing stock, and not a lot of new housing, leading even some companies to build hotels to meet housing needs.

To a question about environmental justice (i.e., for low socio-economic status or minority population) and considerations pertaining to the placement of the wells, Dr. Kelsey replied that much of the drilling is occurring on state forest or game land, or on farms. Many of these regions do not have significant minority populations, but he has not seen a study that examines these issues.

An audience member suggested that noise levels of 60–70 decibels 1,000 feet away are very high and would likely cause elevated cortisol levels and blood pressure—some objective findings that could be identified by research in this area. The person added that during a 30-year work boom it seems clear how one would go about minimizing some of the negatives such as traffic safety and road damage, but the more important question seems to be how to maximize the benefits that arise from the economic boom? Dr. Kelsey replied that this is the central concern of many Pennsylvania communities and landowners who would like to ensure that the dollars that are coming in can be invested for long-term benefit of the community. Dr. Kelsey and colleagues urge local government leaders to ensure that their infrastructure investments have a dual purpose in mind.

An audience member asked about the hurdles to preparing HIA for shale gas extraction, and asked Dr. Witter why her team's HIA, the only one on shale gas extraction to date, was pulled at the last moment. Dr. Witter responded that the process is very political, and that the Colorado School of Public Health researchers were hampered by not having good access to the decision makers, which affected the clarity of their communication. When commissioners were concerned or wanted answers, it was difficult for the HIA team to communicate with them. It was also never made clear to the team what decisions their recommendations would come to bear on, and moreover, information was offered to them in a piecemeal fashion over a long period of time, making it difficult for them to evaluate and reflect on its implications.

A final audience question to Dr. Witter inquired whether she planned to follow up with and study the medical records of people she found had respiratory exposures related to shale gas drilling. The audience member also asked whether the nondisclosure agreements some people sign are proving to be a barrier to obtaining health data. Dr. Witter responded that the questioner's work in Ithaca is farther ahead in looking at individual health effects. The Colorado research team looked at population-level effects and does not have specific plans to move in the direction the questioner indicated.

REFERENCES

Jacquet, J. 2009. Energy boomtowns & natural gas: Implications for Marcellus Shale local governments & rural communities. http://energy.wilkes.edu/PDFFiles/Issues/Energy%20Boomtowns%20and%20Natural%20Gas.pdf (accessed May 30, 2013).

Kelsey, T. W., M. Shields, J. R. Ladlee, and M. Ward. 2011. *Economic impacts of Marcellus Shale in Pennsylvania: Employment and income in 2009.* Williamsport, PA: Pennsylvania College of Technology and Penn State Extension.

MSETC (Marcellus Shale Education & Training Center). 2011. *Pennsylvania statewide Marcellus Shale workforce needs assessment.* Williamsport, PA: Marcellus Shale Education & Training Center.

Rosenman, K. D., M. J. Reilly, and P. K. Henneberger. 2003. Estimating the total number of newly recognized silicosis cases in the United States. *American Journal of Industrial Medicine* 44:141–147.

Witter, R., L. McKenzie, M. Towle, K. Stinson, K. Scott, L. Newman, and J. Adgate. 2011. *Health Impact Assessment for Battlement Mesa, Garfield County Colorado. February 2011 Draft.* Denver, CO: Colorado School of Public Health, University of Colorado Denver.

5

Air Quality

This chapter provides a summary of presentations related to shale gas extraction and implications for air quality. The presentations address how emissions are released during the process, the sources of emissions, the major pollutants that are released, and monitoring efforts. Two presentations discussed specific geographic locations—Texas and Colorado—and their efforts to monitor emissions and assess the potential for health effects. The Colorado presentation describes a health impact assessment (HIA) performed to assess potential risks of unconventional natural gas drilling near a retirement community. The presentations are followed by a summary of the discussion that ensued.

AIR POLLUTION EMISSIONS FROM SHALE GAS DEVELOPMENT AND PRODUCTION

Allen L. Robinson, Ph.D.
Professor, Departments of Mechanical Engineering and Engineering and Public Policy
Carnegie Mellon University

Allen Robinson began the presentation by stating that the processes utilized during shale gas development release emissions that have the potential to affect air quality. In general, when emissions are released into the atmosphere, the concentrations of these emissions are highest in the immediate vicinity of the source and decrease as distance from the source increases. In addition, processes can occur in the atmosphere to chemically transform the pollutants or produce new ones. These processes affect exposure, and the dose and response characteristics lead to the health effect ultimately observed in the population. Limiting the focus solely to the upstream part of this paradigm—the emissions—still necessitates consideration of a plethora of sources when assessing the impacts of a process such as shale gas development. There is an entire production chain to examine, and the release of emissions can occur at every point in the production chain, Dr. Robinson said.

There are many sources of air pollutants along the shale gas development chain. *Hydraulic fracturing* is a single part of the shale gas development operation that occurs for a short period of time. Other activities associated with shale gas development are sources of air pollutants as well. Example of other important sources include

- *site preparation*, including building roads and clearing pads,
- *drilling* the well,
- *truck traffic* to deliver and remove materials and wastes to and from the site,
- *separation and treatment operations* (remove acid gases, remove water from natural gas and *separation* of natural gas from other hydrocarbons),
- *compressor stations* that pressurize natural gas in gathering and transport pipelines,
- *flaring* that burns off contaminated, noncommercial gas,
- *fugitive emissions* that escape unintentionally from cracks or leaks, and
- *blowdown and venting operations*.

The combination of all these activities creates air pollution, and the question that needs to be addressed is what the *net* effects are of these aggregate emissions from all of the activities associated with natural gas development and production on air quality.

Types of Air Pollutants Associated with Hydraulic Fracturing

Natural gas development and production emits criteria pollutants as defined by the Clean Air Act (EPA, 2012). Nitrogen oxides (NO_x) and volatile organic compounds (VOCs) are associated with oil and gas development; in the presence of sunlight, these may react to form ozone and contribute to regional air problems. Regional chemical transport modeling has predicted that ozone may be of particular concern. Nitrogen dioxide and particulate matter ($PM_{2.5}$) emissions are also worrisome, but may be more of a local than a regional issue.

Hazardous air pollutants or air toxics are another category of pollutant that is emitted with shale gas development and production. Many operations necessary for oil and gas development use diesel-powered engines, which emit diesel particulate matter. Furthermore, natural gas-fired engines can be significant sources of formaldehyde, which is also a secondary pollutant. Aromatics (e.g., benzene and toluene) and other VOCs can be released during shale gas production.

Fugitive emissions released during shale gas extraction are also composed of greenhouse gases such as methane. Black carbon from diesel fuel combustion impacts climate. How the benefits of natural gas shift when emissions that directly impact climate change are factored in is an issue to be examined.

Dr. Robinson discussed Figure 5-1, which lists the sources of emissions described above, categorizes them as major or minor sources of pollution, describes the pollutants they emit, and assesses the quality of data. (The values should be considered the "potential to emit" because particulate filters and other technologies, if implemented effectively, can reduce the contributions of these sources.) He noted that there is reasonable understanding of the emissions from diesel-powered engines (e.g., drill rigs, fracturing pumps, and truck traffic). There is less knowledge about emissions from other sources, for example, completion venting, blowdown venting, and fracturing ponds. There may be an increase in knowledge as the field moves to more controlled "green completions" that significantly reduce fugitive emissions of methane and other gases. During the gas production phase, sources such as compressor stations, heaters and dehydrators, condensate tanks, fugitive emissions, and pneumatics emit air pollutants. Some sources have well-defined emission points, and others do not; for most, more data are needed to accurately assess their impacts. Dr. Robinson noted that the main concern is what the net effect is when these pollutants mix in the atmosphere.

Source	NOx	VOC	PM	Air Toxics	Data Quality
		Well development			
Drill Rigs	●	○	●	●	Medium
Frac Pumps	●	○	●	●	Medium
Truck Traffic	●	○	●	●	Medium
Completion Venting		●		●	Poor
Frac ponds		○		?	Poor
		Gas Production			
Compressor Stations	●	●	○	●	Medium
Wellhead compressors	○	○	○	○	Medium
Heaters and dehydrators		○	○	○	Medium
Blowdown venting		○		○	Poor
Condensate Tanks		●		○	Poor
Fugitives		?		○	Poor
Pneumatics		○		○	Poor

● = major source ○ = minor source

FIGURE 5-1 Sources of emissions.
NOTE: NO_x = nitrogen oxide, PM = particulate matter, VOC = volatile organic compound.
SOURCE: Robinson, 2012.

Spatial Scale of Air Pollution

Dr. Robinson commented on the need to consider the spatial scale of air pollution. The air acts as an integrating medium—the emissions from many sources mix to impact air quality and produce exposures. There is much interest in *local* exposure because there are usually homes near gas wells, and emissions from drilling activities that have the potential to affect local residents. But, from an air quality perspective, he stressed, it is imperative to also think at the level of *field* exposure. As well pads become concentrated in a small area, the emissions from the individual activities of each are integrated and can alter air quality. Finally, the *regional* scale needs to be considered. Individual wells may cover a large spatial distribution, and as their numbers increase, they can have a regional impact. This means that NO_x and ozone may not only affect local homes, but may also impact cities that are farther away. For example, Marcellus Shale activities have the potential to affect ozone concentrations in Philadelphia. One way of thinking about regional impacts, Dr. Robinson suggested, is as a massive refinery distributed hundreds or thousands of square miles. From a regional perspective, it is very challenging to fully comprehend the impacts on air quality due to shale gas development.

Another point to consider is where air quality monitoring currently occurs. There is a large network of state and local air monitoring stations that are deployed by state and local agencies to monitor air quality. Selection of sites for stations is population-based, with many stations being located in the Northeast corridor; there is reasonable justification for sitting stations in this fashion. However, there are not many sites in more rural, less populated areas, where much of the gas development is occurring. A monitoring strategy to understand the air quality impacts of shale gas developments in less populated areas, such as in the Marcellus Shale region, is needed. It is possible to make measurements for short-term studies by setting up temporary monitoring stations downwind of well sites. This will indicate very coarsely what the air quality impacts might be. But these types of assessments are sensitive to meteorological and other factors, making it difficult to quantify the true air quality impacts. Furthermore, because of the dearth of permanent monitoring stations in rural areas, there is limited background baseline data, so it is impossible to track the changes in air quality over time. In addition, there is pressure to reduce monitoring stations—not to establish new ones—and this will not aid in ameliorating the spotty nature of the data concerning the impacts of rural shale gas production on air quality.

Predicting Future Trends in Emissions

Dr. Robinson underscored the need to better predict the impact of emissions associated with shale gas production. Carnegie Mellon is developing an emission inventory for chemical transport modeling to predict the regional impacts of Marcellus Shale activities on air quality. The emission inventory covers the states of New York, Pennsylvania, and West Virginia, but the modeling of impacts considers the entire Eastern United States.

In the Marcellus region, new natural gas development has rapidly expanded. There were no Marcellus wells in the mid-2000s. Currently, approximately 1,000 wells are being drilled per year. The future projections are imprecise, but substantial development will certainly continue into the future—perhaps 3,000 wells drilled per year by 2020 (Considine et al., 2010). The associated activities (e.g., drill rigs, hydraulic fracturing pumps) are going to be contributing significant emissions to the area for the foreseeable future. The exact amount depends strongly on the price of natural gas.

Emissions from compressor stations and processing facilities are tied to the amount of gas produced (not necessarily the number of wells drilled). There is a range of estimates of future gas production, which is also expected to increase. By 2020, there could be 15 billion cubic feet per day (bcfd) produced, where currently there is less than 5 bcfd (Considine et al., 2010; U.S. Energy Information Administration, 2013). These activity predictions are important because they factor into the analysis of future emissions.

Dr. Robinson stated that there is also uncertainty in the underlying emissions data. For the diesel-powered sources, there is literature on emissions data and on duty cycles. These data can be used to build reasonably robust estimates. For process-level data or activities such as pneumatics, there is more uncertainty. To estimate these emissions, there are a number of parameters to use. There is the emission factor (which is the rate of production of the pollutant), operations characteristics (such as power), the size of the source, and the amount of time the source is functioning. These values can vary from basin to basin and operator to operator, so they carry some degree of uncertainty, but, despite that challenge, it is possible to account explicitly for that uncertainty in the estimation of emissions.

In a Monte Carlo simulation, probability distributions for factors that have a high degree of uncertainty are used to calculate a range of possible emission levels. For example, ranges of emission factors for drill rigs, hydraulic fracturing pumps, and other sources can be obtained from the literature. There is also increasing amounts of data on how devices are operated in the field. These distributions have a level of uncertainty that can be incorporated into the analysis. Dr. Robinson noted that this type of analysis explicitly acknowledges uncertainty and accounts for it formally, instead of hiding behind it and using it as an excuse for inaction.

These analyses have revealed that aggregate emissions are important. The amount of NO_x emissions due to Marcellus Shale is estimated to be substantial: in 2009, 58 tons per day of NO_x were emitted, and by 2020, this is expected to increase to 129 tons per day. Currently Marcellus sources are responsible for a few percent of the regional NO_x emissions, but, by 2020, this is expected to increase to 12 percent (Roy et al., 2013). Existing regulations are expected to reduce NO_x emissions from power plants and diesel engines. However, as the activities associated with development and production grow; the associated emissions may undermine the gains from regulations and have impacts in the region.

It is possible to do an analogous calculation for VOC emissions. A similar trend is observed—the tonnage of emissions is significant and will increase over time. For VOCs, local effects must be examined as well because the type of gas—wet or dry—varies with location. Wet gas has a higher fraction of condensates and its production releases more VOCs. In the Marcellus region, wet gas is found in West Virginia and southwestern Pennsylvania. The properties of the gas alter its emissions and thus affect the exposure of the surrounding community to air toxics (The Nature Conservancy, 2010).

The analysis also shows that diesel particulate matter and other air toxics may be important components of air pollutants, which again, is predicted to increase. In 2009, approximately 2 tons per day of diesel particulate matter were released, and this is predicted to double by 2020 (Kemball-Cook et al., 2010). Emissions from on-road diesel activity are expected to decrease because diesel particulate filters are now required (EPA, 2006). Combining this regulation with the new, sizeable activity from gas development means that the fraction of emissions due to Marcellus activities is expected to increase.

Regarding ozone, model simulations for the Haynesville Shale in Louisiana and Texas have shown a perturbations as large as 10 parts-per-billion (ppb) in the maximum-daily 8-hour average ozone level for high emission scenarios (Kemball-Cook et al., 2010). Considering that the current air quality standard for ground-level ozone is 75 ppb, this disturbance is a significant increase and an important regional impact.

Dr. Robinson emphasized that effective control measures do exist for emissions from shale gas activities. For example, for compressor stations, technology such as oxidation catalysts, selective catalytic reduction, and stoichiometric combustion engines with three-way catalysts can reduce the emissions of NO_x, VOCs, and formaldehyde. For drill rigs and hydraulic fracturing pumps, it is possible to use selective catalytic reduction and diesel particulate filters to reduce emissions. Fuel switching from diesel to natural gas is also an option, and will reduce diesel particulate matter (but not necessarily NO_x emissions). Finally, removing waste water from drill sites with pipe networks instead of with trucks can reduce emissions. A new oil and gas regulation is going to require green completion, which is predicted to reduce VOCs by 95 percent (EPA, 2006). Flaring also will substantially reduce emissions (by 51 to 84 percent) (EPA, 2006). Vapor recovery units on condensate tanks can reduce emissions by 91 to 95 percent (EPA, 2006). Dr. Robinson noted that society needs to decide what level of emissions is acceptable, and then implement various rules to achieve the necessary reductions. All of the technology currently exists.

When discussing air pollution and natural gas, it cannot be overlooked that natural gas development provides some air quality benefits to end users, Dr. Robinson stated. Natural gas is a cleaner burning fuel and emits less NO_x, sulfur dioxide, PM, and carbon monoxide (Lash and Engelder, 2009). For example, switching from coal to natural gas can improve air quality around power plants. Therefore there are many potential air quality benefits associated with greater use of natural gas. However, one cannot ignore the challenges on the development and production side; it is a matter of deciding what the appropriate level of control is, and then implementing strategies.

AIR QUALITY IMPACTS OF NATURAL GAS OPERATIONS IN TEXAS

Michael Honeycutt, Ph.D.
Director, Toxicology Division
Texas Commission on Environmental Quality

Michael Honeycutt began his presentation by describing several shale formations in Texas, three of which are actively being drilled. The Haynesville Shale is in east Texas; the Barnett Shale is in north Texas; and the Eagle Ford Shale is in south Texas. Because shale gas development can result in air emissions, the Texas Commission on Environmental Quality (TCEQ) has been monitoring these areas for the past 6 years.

Under the Clean Air Act, regulators have categorized air pollutants as either criteria air pollutants or noncriteria air pollutants (EPA, 2012). The criteria pollutants are chemicals with national ambient air quality standards—ozone, particulate matter, lead, SO_x, NO_x, and carbon monoxide. The noncriteria air pollutants are those that fall outside this category.

Dr. Honeycutt noted that there has been exponential growth in gas development within the Barnett Shale over the past decade, but it is now starting to level off. Ozone is the only criteria pollutant associated with oil and gas activities for which the Dallas–Fort Worth (DFW) region is in violation. Despite the exponential growth in the number of wells (from almost zero in 1993 to more than 14,000 in 2011), the 8-hour ozone design value is trending down, from 106 ppb in 1995 to 86 ppb in 2010 (see Figure 5-2) (TCEQ, 2013). (The 8-hour ozone design value for the DFW nonattainment area is 85 ppb.) There was a slight increase in 2011, but that was most likely due to a drought, he suggested.

The Monitoring Process in Practice

Dr. Honeycutt described the types of monitoring that are used to monitor emissions. Observational data on levels of criteria and noncriteria pollutants (such as the instance described above) are collected using fixed-site and short-term monitoring methods. TCEQ carries out two types of long-term fixed-site monitoring. One type uses a fixed-site canister to capture air samples. A stainless steel, clean canister is put under vacuum and attached to a timer and flow controller. On the day that samples will be taken, a valve opens at midnight, and air is pulled through the sampler and into the canister. This continues for a 24-hour period. At midnight the next night, the sampler shuts off. A technician travels to the site the following day, takes out the old canister, and puts in a new one. The sampled canister is shipped to Austin to be analyzed by the toxicology lab. Six days later, the monitor takes another 24-hour sample. This type of monitoring station is permanent and provides a 24-hour sample every sixth day.

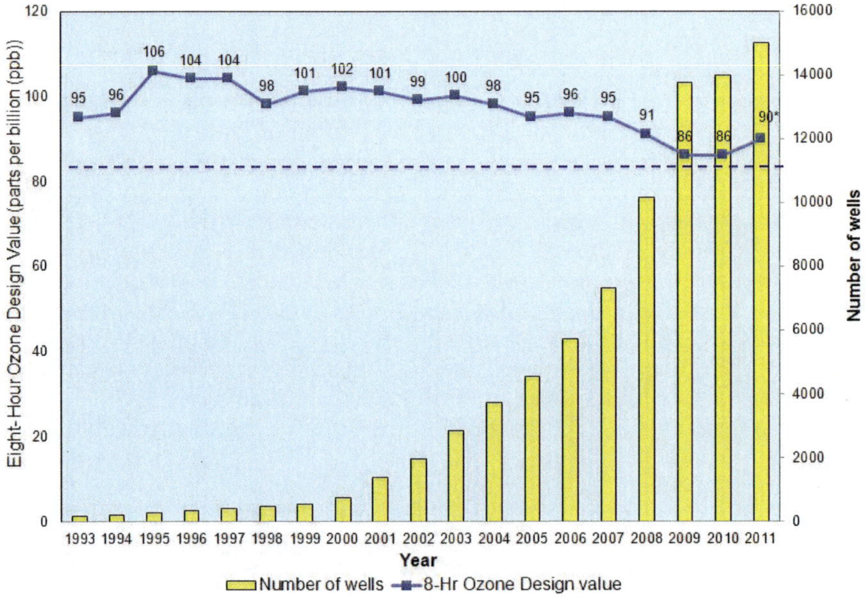

FIGURE 5-2 Dallas–Fort Worth area ozone design values and Barnett Shale production.
SOURCE: Honeycutt, 2012.

The other type of long-term sampling is fixed-site auto gas chromatography (AutoGC). This sampler is housed in a trailer. AutoGC takes an hourly sample; of every hour, 40 minutes are used to collect a continuous air sample, 10 minutes are needed to run the GC analysis, and 10 minutes are required to prepare for the next air sample. This machine operates 24 hours a day and 365 days per year; it collects samples 22 hours per day, 1 hour is devoted to a calibration curve, and the other hour is for a quality assurance/quality control sample. The two control steps are alternated so that the same 2 hours are not omitted each day. Because it provides data continuously, AutoGC provides a wealth of information.

AutoGC can be a fixed-site monitor, but, when necessary, mobile laboratories can be equipped with this technology. These laboratories can be placed right on the property line of oil and gas facilities. Investigators can be out in the field for several weeks at a time collecting samples on many types of chemicals with these monitors.

Dr. Honeycutt reported that a monitor in Longview, Texas, in the Haynesville Shale, detected elevated concentrations of benzene. Benzene is an air toxic that TCEQ is monitoring closely. Benzene is the chemical whose measured concentration is closest to what is considered an acceptable level. Every other chemical that is currently being monitored is further below its acceptable level compared to benzene.

The monitor for this site had been functioning since 1998. Dr. Honeycutt noted that in 2005 there were some slight increases in the annual average benzene concentration, and in 2007 it really climbed. It was above the TCEQ long-term screening value of 1.4 ppb.

This monitor is located 1,100 feet to the northeast of a new oil and gas facility. The facility had only been constructed the year before the monitor started detecting elevated levels of benzene. TCEQ explored the facility and collected air samples. They found that some of the storage tanks were not designed properly. There were significant emissions from those tanks (benzene levels of 1,100 ppb were measured in their vicinity). These benzene leaks were being picked up farther away by the permanent monitoring station. TCEQ worked with the responsible operator to figure out what was going on and to fix it. The next year, benzene levels dropped below 1.4 ppb, and they are still trending downward.

This case underscores a concern of TCEQ about air quality and gas developments. There are many gas developments in Longview, Texas—about one every 0.25 mile. The well immediately south of a monitor was the one to which a problem was traced. TCEQ may have been fortunate to have had a monitor in exactly the right location to pick up on the sole source of benzene emissions in the town. But there was also the possibility that these issues were widespread, and there were no long-term monitors near the rest of the wells to alert TCEQ to problems.

To determine how widespread problems were, TCEQ needed to quickly evaluate many shale gas development wells. To carry out this short-term monitoring and evaluation, TCEQ mounted the GasFindIR™

camera (especially developed for detection of gas leaks) on a helicopter. GasFindIR™ reveals VOC emissions such as methane—they cannot be seen by the naked eye, but, on camera, they appear as dark, billowy smoke. TCEQ has conducted flyovers of the Barnett Shale and Eagle Ford Shale. It costs about $100,000 to conduct a flyover, but it is the best money spent in terms of reducing emissions. Using this camera from a helicopter allows thousands of wells to be surveyed at a time. After the footage is examined, TCEQ can approach the well operators and inform them of a problem. But, finding a problem is atypical. Of the 5,000 storage tanks surveyed in the Barnett Shale, only 88 had significant hydrocarbon emissions.

There are also handheld GasFindIR™ cameras that make it possible to screen many facilities quickly. They are slightly bigger than home camcorders, and field investigators can walk around a facility with them. Since August 2009, TCEQ has surveyed 2,122 gas development sites using GasFindIR™ cameras, and, at 2,078 of these sites, handheld VOC samplers were also used. Whenever there is something of interest, an instantaneous or short-term (30 minutes to 1 hour) canister sample can be taken. On the basis of observations with these instruments, 1,126 canister samples have been collected and analyzed for VOCs. There are other VOCs that TCEQ assesses as well, and this is possible with other instruments that detect nonmethane hydrocarbons. Short-term samples also have been collected and analyzed for carbonyls, NO_x, and sulfur compounds.

Indications from Short-Term Monitoring

Dr. Honeycutt provided an overview of findings from short-term monitoring. He said that less than 5 percent of the more than 1,100 VOC canister samples mentioned above exceeded a short-term, health- or odor-based air monitoring comparison value (AMCV).[1] Texas regulates chemicals for odor, and some VOCs, such as isobutane and isopentane, have been detected above an odor-based AMCV. Some samples (e.g., benzene and carbon disulfide) have been elevated above a short-term health-based AMCV. Most of these were discovered when TCEQ initiated its monitoring program. The actions taken by TCEQ since then have decreased the number of residents' complaints of odor.

TCEQ continues to receive some citizen complaints of odor and irritation, and investigators are experiencing similar symptoms in the field. These complaints are thought to be related to a glycol degradation product. TCEQ researchers are trying to figure out what could possibly be emitted with the aim of designing sampling and analysis techniques to monitor these emissions in the future, Dr. Honeycutt said.

[1] An AMCV is an air quality standard designed to protect health.

Indications from Long-Term Monitoring

Dr. Honeycutt also provided an overview of results from long-term monitoring. Six million to 7 million people live in the DFW metro area (U.S. Census Bureau, 2013). There are about 15,000 wells in the Barnett Shale geologic formation. There are seven fixed-site canister samplers, five fixed-site AutoGC monitors, and two sites where they are co-located (for the purpose of comparing the data between methodologies). There are pros and cons to both types of samplers. The target analyte list for the canister sampler includes about 85 chemicals. The target analyte list for the AutoGC sampler includes only about 55 chemicals. But the hourly aspect of the outputs from the AutoGCs provides more data with which TCEQ can work. The state is prepared to set up five more AutoGC monitors, but it takes time to select an appropriate site for the monitoring station. It is expected that the amount of data available is going to grow exponentially in the near future.

Despite the massive growth in the number of wells, there is no indication from the long-term monitoring of benzene that there is a health risk. (Recall that benzene is the chemical closest to exceeding its acceptable level and is therefore considered the risk driver.) The newer stations are aptly situated both to capture community exposure and to monitor sources. Based on the monitoring data, there is nothing that concerns TCEQ on an airshed level; all levels of benzene are below the 10^{-5} risk level[2] used by the U.S. Environmental Protection Agency (EPA). In fact, some of the lowest monitored levels of benzene across the state are in the Barnett Shale—other parts of Texas have higher benzene concentrations, Dr. Honeycutt said.

In the town of Dish, Texas, many residents were concerned about the effects of shale gas development. The state health department visited the town in 2010 to collect biological samples from 28 residents. The health department examined VOCs in blood and VOC breakdown products in urine, and took samples of tapwater. They compared these data to National Health and Nutrition Examination Survey data and concluded that community-wide exposures from gas wells or compressor stations have not occurred in the sample population.

The results of the analyses of all of TCEQ's samples are online in the form of an interactive map.[3] It is possible to click on a region of the map to determine the concentrations found there, and it is also possible to enter an address to find the data from the nearest station. According to Dr. Honeycutt, this is a great public relations and communications tool.

---

[2] The probability that 1 in 100,000 individuals will contract cancer in a lifetime from continuous exposure to the chemical concentration.

[3] See http://www.tceq.texas.gov/airquality/barnettshale/bshale-viewer (accessed May 30, 2013).

TCEQ's Current Activities

Dr. Honeycutt concluded his remarks by noting that the issues TCEQ did document were found early in the monitoring process. Almost all of those were human and mechanical failures on the part of the companies. Corrective actions for those—tightening valves, replacing gaskets, closing hatches—were easy to implement. TCEQ is still actively executing compliance investigations with handheld monitors, but when an issue is found, there is usually an easy fix for the problem. TCEQ is also implementing new rules for best management practices. It is conducting outreach and attempting to educate well operators and the public. For the well operators, TCEQ instructs them about the differences between percentages and parts per billion as they pertain to monitoring emissions. For the public, they host open houses to disseminate information about shale gas development and monitoring activities. Transparency and citizen access to information is important; therefore all data are available on TCEQ's website.

AIR POLLUTION EXPOSURE AND RISK NEAR UNCONVENTIONAL NATURAL GAS DRILL SITES: AN EXAMPLE FROM GARFIELD COUNTY, COLORADO

John Adgate, Ph.D., M.S.P.H.
Professor and Chair, Department of Environmental and Occupational Health
Colorado School of Public Health

John Adgate began his presentation by describing the term "unconventional natural gas drilling" as the cradle-to-grave process of developing a well and extracting the gas resources. He explained that in response to a citizen request, the Colorado School of Public Health performed a health impact assessment (HIA) on the potential risks of unconventional natural gas drilling (Witter et al., 2011). Natural gas well sites were planned for the area around the retirement community of Battlement Mesa in Garfield County, Colorado. Residents wanted to know what the minimum safe distance was—how close could a drilling site be to their homes and not pose any health risks. HIA is an apt tool for evaluating risk in a situation such as this. Dr. Adgate pointed out that the aim of his presentation was to elaborate on some of the key questions around HIA, and how HIA is used in the public process.

Dr. Adgate provided some context about Garfield County in Colorado. A unique feature of Colorado is that in many locations there is a "split estate," which means that the people who live on the land and the people who possess the mineral rights for the land are often two different parties. A split estate often creates interesting contrasts and pressures when the mineral rights owners want to develop their resources. Part of the conflict in Battlement Mesa, Garfield County, arose from the split estate issue.

There are about 5 million citizens in the state of Colorado and approximately 80 percent of them live east of the Rocky Mountains. Garfield County is in the western part of the state, and has a population of about 50,000. Dr. Adgate noted that in 2010, Garfield County issued 1,806 drilling permits, making it the county that granted the second highest number of permits (Weld County issued the most, at 1,854 permits). Currently, there are more than 8,000 natural gas wells in Garfield County. Across the state, natural gas wells are becoming more common, and development is starting to encroach on populated areas—thus, increasing citizen interest in the effects of drilling on the environment and human health.

Colorado is a very dry state with interesting meteorological patterns. In Garfield County mornings typically bring rising warm air, with subsequent evening cooling leading to downslope winds. Dr. Adgate explained that in Garfield County most of the gas development is occurring along the valley floor, and these weather patterns have implications for what people are exposed to in and around these sites. In addition, the Colorado River bisects Garfield County, and is a source of drinking water for many people. A major interstate also runs through the valley, and it is a source of some of the same air toxics that are associated with natural gas drilling.

The HIA in Garfield County

Dr. Adgate discussed the process used to conduct the HIA. Garfield County has long-term air monitoring stations operated by their Department of Public Health. These stations collected canister air samples over 24-hour periods that were analyzed in EPA-certified labs. The lab quantified 78 nonmethane VOCs. Overall, there were 163 ambient air samples; one was collected every 6 days over the span of almost 3 years. These samples were used to characterize *ambient* air levels in the natural gas development area between 2008 and 2010; this was the period over which production peaked and then started to wane.

To collect additional data for the HIA, representatives of Garfield County Public Health staff visited a well site to conduct air sampling. Surrounding the well site was the supporting infrastructure that develops over time, and farther away were previously established agricultural zones, ranches, and beehives. There were no houses within 1,000 feet of this particular operation. The team noted that hatches were open on the tanks that hold the wastewater after it flows out of the well, and they were able to smell odors.

To characterize the *peak* emissions of the well development process Garfield County Public Health staff collected samples from the period of flowback and well completion while at the well site. The flowback process is when the water used for fracturing exits the well, usually carrying hydrocarbons. There were 24 samples collected at distances between about 150 and 500 feet from these well sites, and they were used as part of this human health risk assessment process.

To develop risk estimates, many questions needed to be addressed. How long are these wells in place? How long does it take to develop a well? How many wells are on a pad? How many pads are there in a community? To address some of these, EPA methods were used to estimate risks over two durations: *subchronic* (short-term effects) and *chronic* (longer-term effects). The subchronic exposure scenario assumed a 20-month period,[4] because this is the length of time a resident would be exposed to pollutants from well development. The chronic exposure scenario assumes a 30-year period because natural gas could be produced from a mature well for 20–30 years.

Dr. Adgate noted that to address uncertainty in these estimations, many assumptions were made during these calculations. One assumption concerned the two distance-related categories of exposure. Exposures were divided into "near" and "far" based on citizen complaints about odor. Essentially, those who could smell gas well activities and lived less than half a mile away were classified as *near* to a well. Everyone else was classified as living *far* from the well. Also, the analysis used "default" options that assume reasonable maximum exposure and are designed to be health protective. They presume that a resident remains in the town 24 hours per day, 350 days per year, which necessitates using the upper bound (95 percent upper confidence limit) of mean concentration estimates for the various air toxics.

To conduct the human health risk assessment, researchers from the Colorado School of Public Health used standard EPA methodology. The researchers combined existing data with interesting new data that had not yet been reported in the scientific literature. The longer-term and peak data just described went into creating exposure scenarios. There were four scenarios: *chronic-far*, *subchronic-far*, *chronic-near*, *subchronic-near*. Both far scenarios relied only on the ambient air data from the permanent monitoring stations. The subchronic-near scenario used only the 24 air samples from the peak well-completion phase. The chronic-near scenario used the time-weighted average of the peak well-completion samples (a duration of 20 months) and ambient air samples over a duration of 30 years.

---

[4] In Garfield County, 9 months to 1 year are required to develop a well, a well may undergo hydraulic fracturing multiple times, and there are usually multiple wells in a pad. Thus, the well development process may take up to 20 months or more.

To quantify noncancer health risks, hazard indexes (HIs) were used. The HI is the sum of all applicable semiquantitative hazard quotients (HQs). For each compound detected, an HQ is computed, which is the ratio of the estimated exposure to the reference concentration (RfC). RfCs are values that are suggestive of exposures at which noncancer health effects may occur, and are catalogued in the EPA's Integrated Risk Information System. This system provides both chronic and subchronic RfCs, and each RfC was applied in the appropriate scenario. Dr. Adgate said that the meaningful question is whether the HI is greater than 1. If it is less than 1, it is possible to be confident of few health effects. If it is greater than 1, then risks to health are more likely.

The process of estimating cancer risk is distinct from estimating noncancer health effects. The long-term average exposure for each carcinogen was estimated for a 30-year time period using the ambient air sampling data. This exposure was multiplied by an inhalation unit risk value, which indicates potency of the carcinogen, to compute the lifetime excess cancer risk. The risk level was summed across the different carcinogens that were detected to compute cumulative lifetime cancer risk. The lifetime cancer risk is an indication of the number of excess cancers in a population of 1 million people. The criterion used to evaluate risk from carcinogens is whether the risks are greater than 1 in 1 million (i.e., is there more than 1 excess cancer within a population of 1 million people).

Findings of the HIA

Dr. Adgate reiterated that there were four different scenarios used to calculate HIs: chronic-far, subchronic-far, chronic-near, subchronic-near. For noncancer health effects, it was determined that the HI is greater for those living closer to well sites (see Figure 5-3), although most HIs did not reach a level of concern. The HI for chronic-far was 0.4, and the HI for chronic-near was 1. The HI for subchronic-far was 0.2, and the HI for subchronic-near was 5 (McKenzie et al., 2012). Only the HI for subchronic-near was above 1—which is above the level of concern. The next question is what can be done about this, that is, how can exposures and risk be reduced.

Instead of averaging all HQs to find a single HI, it is possible to categorize toxics by health outcome and calculate an HI for neurological effects only, or respiratory effects only, etc. For the subchronic-near exposures, the data were parsed on the basis of health end points. For neurological effects, the HI was 4. For respiratory effects, the HI was 2. For hematological effects, the HI was 3. For developmental effects, the HI was 1 (see Figure 5-4) (McKenzie et al., 2012). The first three health end point categories are in the range of concern, Dr. Adgate noted, where action likely should be taken to reduce exposure. For example, new controls could be mandated to be put in place to limit citizens' exposures.

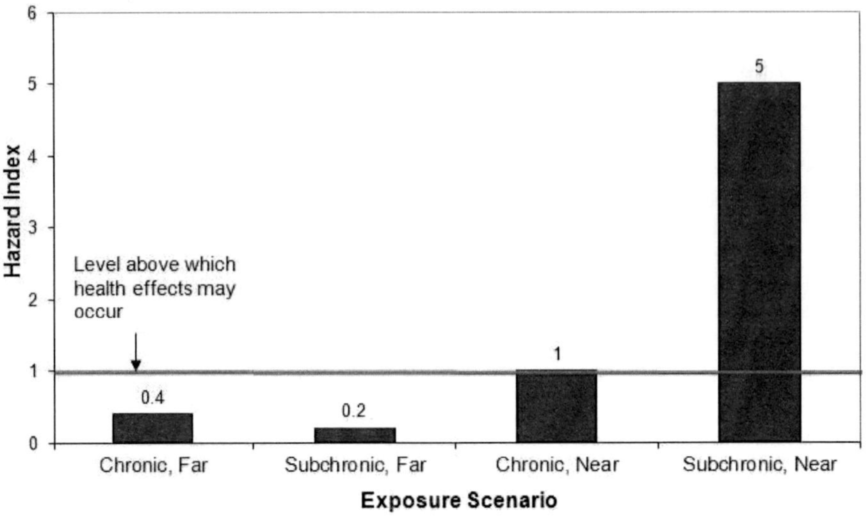

FIGURE 5-3 Hazard indexes by duration of exposure and distance from source.
SOURCE: Adgate, 2012.

The biggest contributor to the hazard index was trimethylbenzenes, which were responsible for 46 percent of the estimated risk. The next biggest contributors were aliphatic hydrocarbons (21 percent) and xylenes (15 percent) (McKenzie et al., 2012). Adgate noted that benzene was a relatively small contributor, at only 5 percent. So, a range of compounds is responsible for the health effects, and a fair amount is known about these from a toxicological standpoint. These data may inform some of the practices aimed at reducing risks and health effects.

Dr. Adgate noted that the estimated cancer risks for a 30-year exposure were in the range that does not typically warrant concern. There was a 6 in 1 million risk for residents living far from a well, and 10 in 1 million risk for residents living close to a well. This is above the 1 in 1 million risk level, but below the level at which the EPA typically requires remediation. Overall, concentrations were similar to or higher than those observed in many urban areas. Benzene, although it contributed little to the noncancer health effects, was, along with ethylbenzene, the main risk driver for cancer risk.

Residents living near well completion activities are potentially exposed to substantial levels of air toxics. Subchronic noncancer cumulative and end point specific hazard indexes are greater than 1 for residents living near well pads. Also, estimated cancer risks and chronic noncancer hazard indexes are greater for these residents living near the well pads, but are within a generally acceptable range.

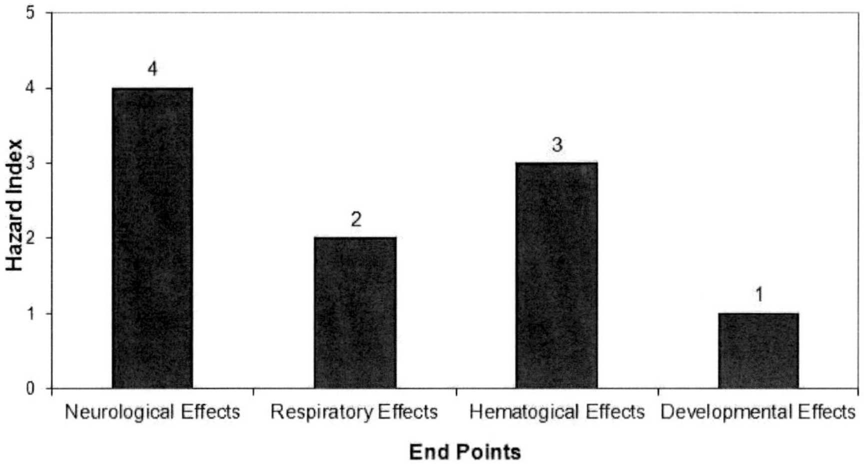

FIGURE 5-4 Hazard indexes by health end point: Near wells, 20-month exposure scenario.
SOURCE: Adgate, 2012.

There are many uncertainties and limitations within this study, Dr. Adgate noted. The sample size was small, using only 24 samples to assess 78 compounds collected from 150 to 500 feet around the well source. In general, limited data exist on emissions around well completion sites. In addition, nonmethane pollutant emissions appear to vary substantially by field type, number of wellheads, completion process, and controls used. Only a limited suite of volatile compounds was explored here, which excluded other primary or secondary pollutants (e.g., aldehydes, diesel exhaust, etc.). In summary, Dr. Adgate said that short-term exposures are probably important to address and that prevention strategies should be directed at minimizing exposures during well completion activities to reduce potential subchronic noncancer risks.

AIR QUALITY RESPONDENT

Bernard D. Goldstein, M.D.
Professor Emeritus, Department of Environmental and Occupational Health
Graduate School of Public Health, University of Pittsburgh

Dr. Goldstein began his commentary by noting that in southwestern Pennsylvania, Shell is constructing a plant to manufacture ethylene through the process of cracking, or heating hydrocarbons to induce unsaturation. It is currently in the middle of the permitting process.

During this process, the plant's potential for emitting ozone precursors will be examined.

He noted that Dr. Robinson's point about emissions being an aggregate issue is extremely relevant to the southwestern Pennsylvania case. This one plant, which will have to go through all of the ozone considerations in a region that has borderline-elevated ozone levels may produce fewer emissions than all of the Marcellus Shale wells combined. But each individual Marcellus Shale well does not have to go through the ozone permitting process because it is a small source of emissions. There is an "emissions threshold," and below this threshold, no regulation is required.

This threshold also exists for the Emergency Planning and Community Right-to-Know Act (Executive Order No. 12,969). This law states that reports must be made to the community if the amount of a chemical stored in or released from a factory or other such structure exceeds a certain limit. This is especially relevant for emergency management, so that firemen and other first responders can be educated about the risks of these chemicals in case there is ever a major problem. But not one of the individual Marcellus Shale sites exceeds the threshold to initiate this kind of communication or training. Again, that is an issue of the aggregate versus the individual, which Dr. Robinson laid out well.

Ozone and formaldehyde are two key examples of emissions that result from natural gas development and use, and that may affect health. According to Dr. Goldstein, the ozone issue has a particular irony to it, in that if the ozone standard is exceeded, or an area starts to approach nonattainment, a state implementation plan needs to be developed. To meet the ozone standard, this plan will likely mandate restriction of some industrial development. The irony is, of course, that the Marcellus Shale is being sold to Pennsylvanian communities as, if you will, industrial development.

Formaldehyde is another interesting issue. Formaldehyde is a one-carbon aldehyde. Methane is a one-carbon hydrocarbon. Burning methane is pretty clean. The one thing of concern with incomplete combustion of methane is formaldehyde.

Regarding emissions of pollutants such as ozone and formaldehyde, what the public health community is requesting is access to information so that health studies can begin. The public health community is concerned about an impact of shale gas drilling on health. It is recognized that there may be an impact on the industry due to the health community's evaluation of these compounds' potential adverse health effects.

Current research into ozone health effects is likely to reduce the ozone standard because of findings relating ozone exposure to premature mortality. There is a similar battle over formaldehyde. Based on recent research it appears that formaldehyde induces leukemia in humans. Research on formaldehyde will affect how natural gas is used, but it needs to be thought of as a bidirectional issue, and the public health community needs to continue to communicate—to advocate for controls that protect health.

Finally, Dr. Goldstein emphasized, it is important to make a distinction between the two types of air pollutants that have been discussed. One kind includes pollutants that no one wants in the air, that is, NO_x, ozone, and formaldehyde. The other type includes pollutant that industry can sell, be it methane or benzene. What is evident is that, over time, industry does a better job recovering products that they can then sell in the market, particularly if companies are pushed to emit less. So, a distinction between these two types of pollutants needs to be made in terms of strategies to reduce emissions.

DISCUSSION

During the discussion period a question was raised about fugitive methane and motivations to reduce it. Dr. Robinson responded that there are motivations to try to reduce fugitive methane. He noted that fugitive methane is associated with what is called completion venting, which is one of the last stages in bringing a well online. Very high estimates of fugitive methane imply that there is venting into the atmosphere and steps to be more environmentally friendly are not being taken. Estimates of fugitive methane are unclear, but are said to range as high as 7 percent; however, 2 percent is about the level where people begin to become concerned about the impact on climate. Climate is the motivator to reduce fugitive methane.

Another audience member raised a question about geographic air pollution and monitoring. The participant stated that NO_x, VOCs, and methane are not ozone but atmospherically transform into ozone. There is some research that has looked at how NO_x can scavenge ozone where it is being released, and that ozone concentrations are not necessarily highest where these ozone precursors are emitted. The participant asked the panel whether this fact had any implications for where to measure ozone. Dr. Robinson responded that this phenomenon is known as the NO_x titration effect, where weekend levels of ozone are higher than on weekdays despite the fact that emissions are typically lower on weekends. This phenomenon will change the spatial pattern of what is going on, and so it is important to keep that in mind. Dr. Robinson also pointed out that considering secondary pollutant formation is also important. This depends on the chemical state of the atmosphere, that is, how much NO_x there is relative to ozone. The effect of additional NO_x will depend on geography and sensitivity. Modeling suggests that the northeast United States is more NO_x sensitive and that additional NO_x would have a bigger impact. Other regions of the country may be more

VOC sensitive. So this factor should be taken into consideration when setting emission controls.

REFERENCES

Adgate, J. 2012. *Air pollution exposure and risk near unconventional natural gas drill sites: Examples from Garfield County, Colorado*. PowerPoint presentation at the Institute of Medicine workshop on the Health Impact Assessment of New Energy Sources: Shale Gas Extraction, Washington, DC.

Considine, T. J., R. Watson, and S. Blumsack. 2010. The economic impacts of the Pennsylvania Marcellus Shale natural gas play: An update. The Pennsylvania State University, College of Earth and Mineral Sciences, Department of Energy and Mineral Engineering. http://marcelluscoalition.org/wp-content/uploads/2010/05/PA-Marcellus-Updated-Economic-Impacts-5.24.10.3.pdf (accessed May 30, 2013).

EPA (U.S. Environmental Protection Agency). 2006. *Diesel retrofit technology: An analysis of the cost-effectiveness of reducing particulate matter emissions from heavy-duty diesel engines through retrofits*. Washington, DC: Office of Transportation and Air Quality.

EPA. 2012. History of the Clean Air Act. http://epa.gov/oar/caa/caa_history.html#caa70 (accessed May 30, 2013).

Honeycutt, M. 2012. *Air quality impacts of natural gas operations in Texas*. PowerPoint presentation at the Institute of Medicine workshop on the Health Impact Assessment of New Energy Sources: Shale Gas Extraction, Washington, DC.

Kemball-Cook, S., A. Bar-Ilan, J. Grant, L. Parker, J. Jung, W. Santamaria, J. Mathews, and G. Yarwood. 2010. Ozone impacts of natural gas development in the Haynesville Shale. *Environmental Science & Technology* 44(24):9357–9363.

Lash, G. G., and T. Engelder. 2009. Tracking the burial and tectonic history of Devonian shale of the Appalachian Basin by analysis of joint intersection style. *Geological Society of America Bulletin* 121:265–277.

McKenzie, L. M., R. Z. Witter, L. S. Newman, and J. L. Adgate. 2012. Human health risk assessment of air emissions from development of unconventional natural gas resources. *Science of the Total Environment* 424:79–87.

The Nature Conservancy. 2010. *The Nature Conservancy's roots of innovation 2010 annual report*. Arlington, VA: The Nature Conservancy.

Robinson, A. L. 2012. *Air pollutant emissions from shale gas development and production*. PowerPoint presentation at the Institute of Medicine workshop on the Health Impact Assessment of New Energy Sources: Shale Gas Extraction, Washington, DC.

Roy, A. A., P. J. Adams, and A. L. Robinson. 2013. Air pollutant emissions from the development, production, and processing of Marcellus Shale natural gas. *Journal of the Air & Waste Management Association.* doi: 10.1080/10962247.2013.826151.

TCEQ (Texas Commission on Environmental Quality). 2013. Ozone design values for the DFW area: Comparison of ozone design values for the Dallas-Fort Worth area for the years 1991 through 2009. http://www.tceq.state.tx.us/publications/pd/020/10-02/dfw-alt-text (accessed May 30, 2013).

U.S. Census Bureau. 2013. State & county quickfacts: Dallas, Texas. http://quickfacts.census.gov/qfd/states/48/4819000.html (accessed May 30, 2013).

U.S. Energy Information Administration. 2013. *Annual energy outlook 2013 with projections to 2040.* Washington, DC: U.S. Energy Information Administration.

Witter, R., L. McKenzie, M. Towle, K. Stinson, K. Scott, L. Newman, and J. Adgate. 2011. *Health Impact Assessment for Battlement Mesa, Garfield County Colorado. February 2011 Draft.* Denver, CO: Colorado School of Public Health, University of Colorado Denver.

6

Water Quality

This chapter provides a summary of presentations that describe the use of water in the process of hydraulic fracturing and the potential impact of shale gas extraction on water resources and human health. Health impact assessment (HIA) is discussed as a framework for assessing the impact of hydraulic fracturing on water. Studies to address public concerns regarding water contamination are also described.

POTENTIAL IMPACTS OF HYDRAULIC FRACTURING ON WATER RESOURCES

Deborah L. Swackhamer, Ph.D., M.S.
Professor, Co-Director of the Water Resources Center
Division of Environmental Health Sciences
University of Minnesota School of Public Health

Deborah L. Swackhamer described her presentation as setting the stage for a discussion of hydraulic fracturing and its impact on water resources. Water contamination and the impact that hydraulic fracturing has on water resources is a growing concern to the public. Dr. Swackhamer's presentation provides the context of the role water plays in hydraulic fracturing and the interaction between hydraulic fracturing and water resources. For this purpose it is important to start by understanding the water cycle in the fracturing process.

There are six important steps in the cycle as described below and seen in Figure 6-1.

1. Water acquisition. Water needs to be collected from a major water site. It could be groundwater or surface water. The water then needs to be transported to the well site.
2. Chemical mixing. Water needs to be mixed with chemicals necessary for the gas extraction. In most occasions it is done onsite or transported as the fluid to be injected.

3. Well injection. The fluid is then injected into the well to fracture the shale and extract the gas. The fluid displaces the gas from the fissures of the rock to be collected.
4. Flowback and produced water. Some of the fluid flows back after being injected into the well. It has been calculated that about 40 percent of the flowback is recovered but varies considerably from well to well. Produced water is that which is recovered with the gas extraction. The flowback and produced water have high concentrations of sand that were mixed during the injection, contaminants from the geological formation, chemical additives, and high dissolved solids.
5. Storage tanks and pit. Most of the produced water is stored in tanks or in an open pit. The fluid is stored before it is treated or disposed. It has a high concentration of chemicals and particles.
6. Water disposal. The water is transported to a treatment facility or disposed by deep well injection, or treated on site.

There is a potential problem for contamination in each one of the steps of the cycle. Dr. Swackhamer noted that the U.S. Environmental Protection Agency (EPA) is carrying out a study to specifically examine impact of hydraulic fracturing on drinking water resources (EPA, 2011). In addition to the fracturing water cycle, it is also important to consider the entire life cycle of water included in the processes of well construction, sand mining, water acquisition, and treatment and disposal during shale gas extraction, Dr. Swackhammer said.

She further explained that during well construction (see Figure 6-2) there are a variety of liquids and muds involved in the process. These are not comparable to the total usage of water during the life of the well, but can constitute approximately half a million gallons and need to be taken into account. The drilling of the well also involves the use of chemicals, so there is a potential for spillage and contamination. For the purpose of gas extraction, the well is drilled several thousand feet past aquifers and surface waters. The higher risk during this process is leakage from the casing. The cuttings, result of the drilling, should also be collected and properly disposed.

Water Use

Sand mining raises a number of concerns that go beyond water. Sand mining requires considerable amounts of water for cleaning and sorting the sand, and in preparing it to be used as a proppant. In addition to the large quantities of water necessary, sand mining raises concerns related to transportation, air quality, workers health, and safety. In the last decade the number of wells has increased significantly making the demand for sand and water rise exponentially.

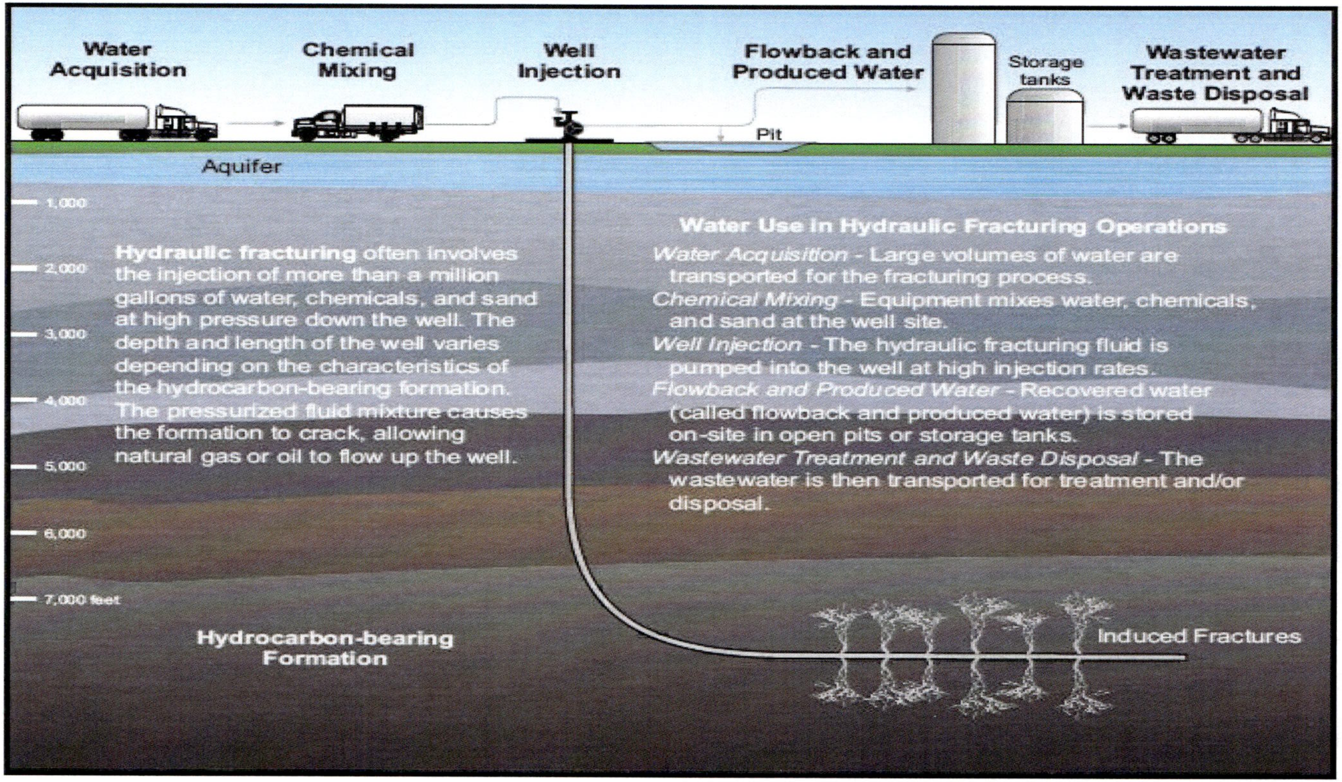

FIGURE 6-1 Water cycle of hydraulic fracturing.
SOURCE: EPA, 2011.

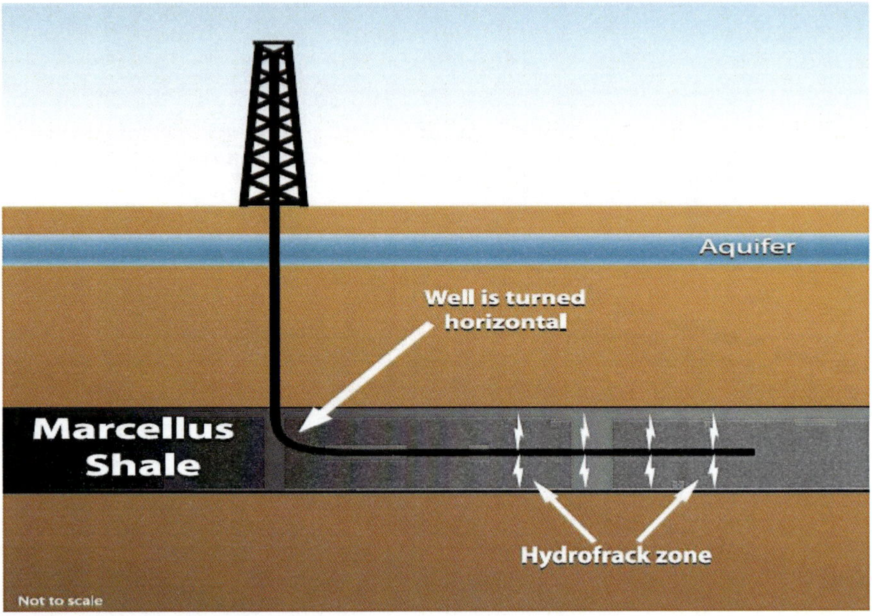

FIGURE 6-2 Initial well drilling and design of a well.
SOURCE: Laurie Barr. Reprinted with permission from Shutterstock.

It is estimated that between 2 and 4 million gallons of water are used in the lifetime of a shale gas extraction well. There are tens of thousands, almost approaching hundreds of thousands of wells being drilled and in production thus a significant amount of water is required in the activity of hydraulic fracturing. The main concern of local governments affected by hydraulic fracturing is the balance between water withdrawal and water consumption. It is important to keep in mind all activities and services provided by the water source. Dr. Swackhamer said that many water sources are being depleted and other sources will be exhausted if the withdrawal rate continues expected in the projection estimated for the Barnett Shale (see Figure 6-3). The graph in Figure 6-4 describes the increase in the number of active wells relative to the required amount of water required for shale gas extraction.

Dr. Swackhamer noted that the impacts of withdrawal of large quantities of water are many. Initially groundwater–surface–water interactions will be affected; excessive withdrawals of groundwater can result in reduced surface water flow or the drying of streams. Groundwater withdrawals can also have huge impacts on ecological functioning. The impact of water withdrawal can be short term or long term; for example, as water changes place underground, exchanges between aquifers can occur. It is imperative to understand the full water balance.

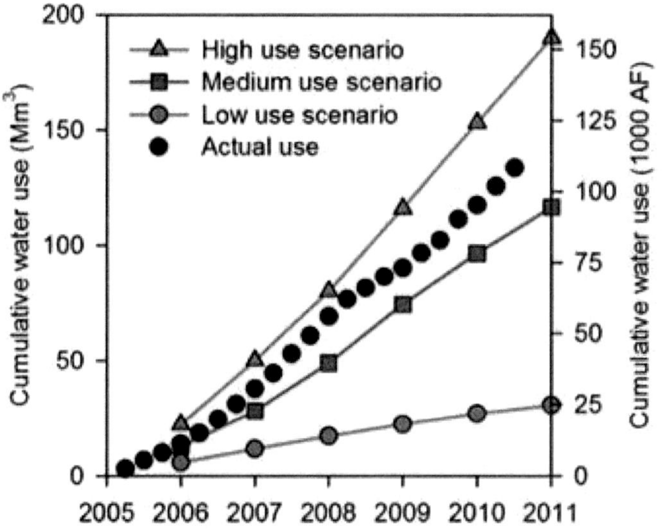

FIGURE 6-3 Postaudit analysis of water-use projections (solid lines) made in 2006 relative to actual water use (dots) through mid-2011 for the Barnett Shale (cumulative as of June 2011).
NOTE: This figure gives an estimate of the uncertainty associated with the analysis, which provides cumulative water use projections within less than a factor of 2 in the next 5–10 years. The assumption that current trends will still be valid beyond the 10-year horizon becomes weaker with increased uncertainty in the projections. Postaudits of long-term projections show that they often deviate from estimates because of unpredicted events, with unprecedented water-intensive shale-gas production being an example.
SOURCE: Nicot and Scanlon, 2012. Reprinted with permission. Copyright © 2012 American Chemical Society.

Water Treatment and Disposal

Dr. Swackhamer stated that the large consumption of water consequently has raised concerns in local health departments regarding water disposal and treatment. There are no studies that could provide a baseline of the particle and chemical concentrations in the water before and after treatment. Some localities use wastewater treatment plants to dispose of the water, but it has been said that not all wastewater treatment plants are capable of treating all chemicals found in the wastewater from fracturing. Wastewater treatment plants were not designed to treat for some of those contaminants. Until recently, Pennsylvania

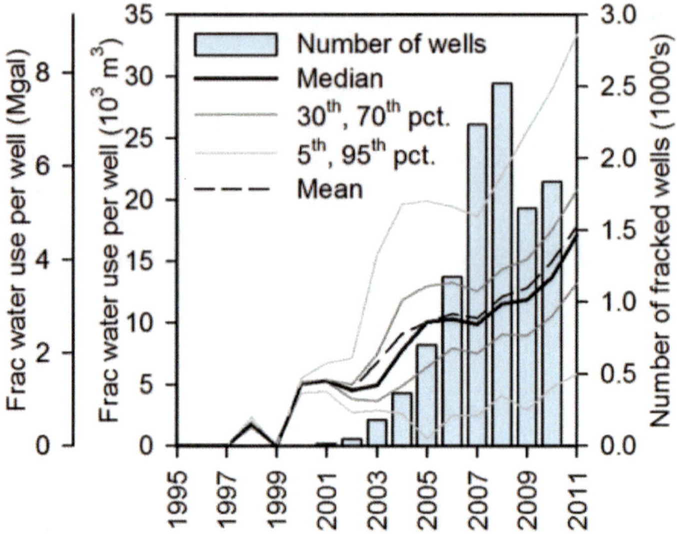

FIGURE 6-4 Time evolution of Barnett Shale well count and water use per well percentiles.
SOURCE: Nicot and Scanlon, 2012. Reprinted with permission. Copyright © 2012 American Chemical Society.

is one of the states that allow wastewater treatment plants to collect the wastewater. This practice is of great concern because there have been a number of studies that demonstrate the fate of many of the contaminants is unknown. The current situation is that either they are not on the safe drinking water list or they are not on the list of what is being measured for wastewater treatment.

Many of the well sites also use a pit to collect their wastewater. Pits are used for temporary storage and to control contamination. Limited treatment is done in these pits that could accumulate high concentration of chemicals and residue from the fracturing activity. A major concern with the pits is a possible overflow or spill. Pits are exposed to the environment and are vulnerable to temperature and weather conditions. The common practice is to transport this water to a different location for disposal. Some wells have developed the capacity to reuse the water or recycle it and inject it in the well once again. After a certain period of time, somewhere between 5 and 7 years, companies plan to refracture wells to increase the productivity. In this there is a potential for cumulative impacts. There are no studies conducted on this practice. The contaminant levels are not known in any step of the cycle and each time the water is reused there is a higher degree of unknowns. The repressurization of the same well structures increases the risk for leaks to occur as well.

Concentrations of the chemicals in the fluid constitute approximately 1 to 2 percent of the total makeup of the fracturing fluid. While 1 or 2 percent of hydraulic fracturing fluid appears very small, Swackhamer noted that the minimum contaminant levels for some of these chemicals in drinking water are far below a total percentage point. Some of the chemicals used in hydraulic fracturing are tested on a scale of parts per million and even parts per billion, which is less than 0.0001 percent. Dr. Swackhamer stated that some of the chemicals used are known endocrine disruptors; these chemicals are measured on a scale of parts per trillion in the environment.

Potential Risks

Dr. Swackhamer stressed that each of the steps of the water cycle in hydraulic fracturing constitutes a potential risk for water contamination and consequent impact on the environment. Not only with wells that are used for shale gas extraction, but in general all wells of similar construction present the most noticeable weaknesses in their physical structure. Most reports on contamination of aquifers pinpoint the cause of contamination as a leak of the well casing, pipes, or storage tanks. The integrity of the wells is critical in their operation and in the minimization of risk. Some of the wells are intended to be used several times, debilitating the structure and compromising operations. In terms of water use, the fluid is injected in the well at extremely high pressures under circumstances that are not completely known. Any engineering structure has a failure rate.

After the fluid has been injected and the fissures in the rock have been opened, there is not a clear indication of what happens to that water. There are estimates that the flowback or collected water is around 40 percent, but some locations have reported a 20 percent and even 80 percent flowback. Considering that a large percentage of the water remains in the ground, it is possible that it can migrate upward or continue to flow for long distances. The gases themselves can be pushed and potentially contaminate aquifers. Another consequence of not accounting for the total balance of injected water and flowback is the accumulation of chemicals in the ground. All the chemicals that are injected in the fluid could prove a more serious long-term contamination problem. Even though the injection is done several thousand feet underground, the composition of the soil is being altered and there is a disruption in the ecosystem, with unknown consequences.

As mentioned before, the treatment and disposal of the wastewater is the toughest challenge. It is a risk to dispose of it in existing wastewater treatment plants because some of the chemicals are not accounted for in the treatment process. There is also a high risk to use injection wells for disposal because the concentration of chemicals can affect other aquifers. There are suggestions for land use application of wastewater treatment solids, but there are many unknowns with this practice and other areas could be exposed to the contamination.

Health Impact Assessment

HIAs, Dr. Swackhamer suggested, are an excellent framework for assessing system effects on human health. Different from a risk assessment, HIA is a more flexible framework that lends itself more appropriately for system-based issues. The difficulty in implementing HIA, is that in the third step, when an assessment is conducted, there are huge data gaps. The initial baseline data for tens of thousands of wells have not been collected. This is a large obstacle when trying to understand the impact on the communities, the environment, and health. Added to this, is lack of coordination or standardization of the data that are being reported. Further, there are many gray lines between the authorities (federal, state, local) that need to regulate hydraulic fracturing. It has been mentioned that the local government needs to become involved in the regulatory process. HIA includes an evaluation of alternatives and nothing has been developed in this area.

Dr. Swackhamer noted there is a call for industry and research institutions to collaborate especially around knowledge gaps. These areas include understanding the

- fate of the fracturing fluid;
- toxic burden for exposure analysis;
- impact of flowback water and produced water;
- effectiveness of contaminant removal and disposal technologies;
- cumulative impacts of refracturing;
- research alternatives to the current hydraulic fracturing practices;
- effective monitoring strategies to be implemented;
- fingerprints of chemicals and fracturing fluids;
- exposure modeling of populations, including vulnerable populations; and
- social impacts and outcomes.

There are many knowledge gaps and much research is needed, she said.

Dr. Swackhamer pointed out that there is a tremendous need for research around toxicity and risks associated with chemical constituents and fluids. A recent paper Colborn et al. (2011) identified more than 600 chemical constituents and fracturing fluids used in the process. Of these, the author could evaluate the literature for the potential health effects of 353 chemicals identified by Chemical Abstract Services (CAS) numbers (see Figure 6-5). The authors found that more than 75 percent of these chemicals can cause acute effects (see Table 6-1).

Dr. Swackhamer closed her presentation by acknowledging that there is much left to do to better understand the potential impacts of hydraulic fracturing on water resources.

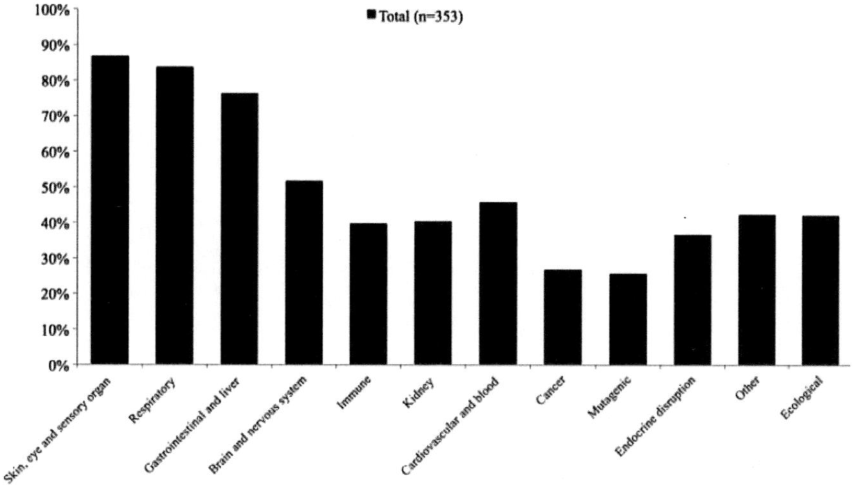

FIGURE 6-5 Profile of possible health effects of chemicals with Chemical Abstract Service (CAS) numbers used in natural gas operations.
NOTE: The x-axis refers to the 12 possible health effect categories and the y-axis represents the percentage of the 353 chemicals with CAS numbers that are associated with each health effect category. The labels on the x-axis are as follows (from left to right): skin, eye, and sensory organ; respiratory; gastrointestinal and liver; brain and nervous system; immune; kidney; cardiovascular and blood; cancer; mutagenic; endocrine disruption; other; and ecological.
SOURCE: Colborn et al., 2011. Used with permission from Taylor & Francis.

TABLE 6-1 Percent of Chemicals in Fracturing Fluids Identified in Colborn et al. (2011) That Could Present Health Impacts

% of Chemicals	Health Impacts
>75	Could affect skin, eyes, other sensory organs, and respiratory and gastrointestinal systems
40–50	Could affect brain and nervous systems, immune system, cardiovascular systems, kidneys
37	Could affect endocrine system
25	Could cause cancer and mutations

SOURCE: Swackhamer, 2012.

HYDRAULIC FRACTURING, WATER RESOURCES, AND HUMAN HEALTH

Robert B. Jackson, Ph.D., M.S.
Nicholas Chair of Global Environmental Change
Nicholas School of the Environment
Professor, Department of Biology
Duke University

Robert Jackson began the presentation by acknowledging his collaborator, Avner Vengosh at Duke University. He stated that the goal of the work he was about to describe was conducted to help answer many of the questions about the use of water in hydraulic fracturing activities. The challenge is not only to collect the data, but more importantly to know which kind of information needs to be produced.

Public concerns about the role of water in hydraulic fracturing have helped to shape the study. The public has voiced many concerns: What is the drinking water contamination potential? What is the amount of water required for the operations? How is the wastewater going to be disposed? What is the concentration of polluting chemicals? These are important questions that need answers.

Dr. Jackson explained the possible water interactions in the hydraulic fracturing process. There are several operations occurring at the surface level, there is injection of fluid far beneath aquifers, and there is the interaction with produced water.[1] Water interactions can occur at different levels. There is natural water or formation water[2] deep underground which often contains high concentrations of naturally occurring chemicals and contaminants and they typically are very salty. In some areas of the country, such as the Marcellus Shale, formation water can contain naturally occurring radioactive materials (NORMs). In hydraulic fracturing, some of the formation waters can flow back to the surface as part of produced waters. Those waters should be kept from contaminating the surface groundwater where drinking water is obtained.

Fracturing fluids used in the hydraulic fracturing process interact with water sources. These fluids represent a small component of water, about 1 percent. But the average operation uses 3 million or 4 million gallons of water, or 30 million pounds, or about 300,000 pounds of fracturing fluid chemicals. Most of those chemicals are harmless such as salt, and citric acid. However, some of them are not as harmless: benzene, naphthalene, and diesel, which are potential carcinogens; toluene and hydrochloric acid and many other hazardous air pollutants; and many

---

[1] "Produced water" is used in the oil industry to describe water that is produced when oil and gas are extracted from the ground.
[2] Formation water is a natural water layer underlying oil and gas reservoirs.

other chemicals, including 2-butoxyethanol, ethylene glycol, and lead (U.S. House of Representatives, 2011). A 2011 report on constituents of fracturing fluids identified 2,500 fracturing products containing 750 chemicals and other components. The presence of a chemical in the field does not mean that it is a problem, Dr. Jackson explained, but it is important to know the concentration, how it got there, and its long-term impact in the environment.

Dr. Jackson paused to highlight a novel approach to chemicals and the hydraulic fracturing process. The chief executive officer of an unconventional gas exploration company in Northern Ireland promulgated a no chemicals pledge for their hydraulic fracturing process (see Box 6-1). Such an approach has not been taken in the United States.

Management of Produced Water

Dr. Jackson turned to discuss produced waters which are primarily a combination of naturally occurring deep formation waters and fracturing fluids. In the Marcellus Shale formation and many other locations, formation waters are very salty, and they may have high bromide concentrations. Bromide can be an issue if it interacts with other chemicals; for example, it can enhance disinfection by-products (e.g., trihalomethanes) upon chlorination of downstream potable water. There are also high concentrations of other toxic elements, including barium, arsenic, selenium, and lead. Hydrocarbon residuals in produced waters, such as oil and organics, can come from both the natural formations and the fluids themselves.

Dr. Jackson reiterated that questions the public would like answers to are the long-term ecological effects and health risks associated with produced water disposal. He described the five main practices used by industry to dispose and manage produced waters. Deep injection for underground disposal is a common practice. The potential problem with deep injection is well leakage or contamination of surrounding aquifers, but there is a long history of this practice, Dr. Jackson stated.

BOX 6-1
No Chemicals Pledge

The Chief Executive of Tamboran, Richard Moorman, came out with a no chemicals pledge for their hydraulic fracturing in Ireland. He said: “Tamboran will not utilize any chemicals in its hydraulic fracturing process in Northern Ireland, and we will be bringing together the best technologies developed worldwide into this one project to ensure the safe and responsible development of a tremendous resource.”

SOURCE: Moorman, 2010.

Another disposal strategy is spraying produced water on land. This strategy is problematic for a number of reasons. As noted earlier, the salinity of produced water is high. The potential for long-term damage of the soil is also high. Further, runoff from sprayed water could contaminate surrounding surface water or percolate into aquifers. Dr. Jackson opined that from every point of view, this strategy is problematic.

The delivery of produced waters to a municipal wastewater treatment plant is another option. Some states have established this practice. It is not recommended to dispose of produced waters in municipal wastewater treatment plants, Dr. Jackson said. These plants do not have the capacity to treat many of the chemicals found in produced waters. Further, some plants do not have the capacity to monitor some of the contaminants. Wastewater treatment plants may then dispose of the treated water in adjacent streams or rivers; the potential for contamination of those water sources is high. The chemicals could accumulate in sediments and cause a higher environmental impact. Municipalities that are located downstream will use the surface water, and those chemicals could end in their drinking water. Figure 6-6 describes the downstream flow of chlorine from the outflow area of a treatment facility. Different concentrations of bromide and trace metals as well as radionuclides in river sediments were found at distances from 300 to 500 meters downstream.

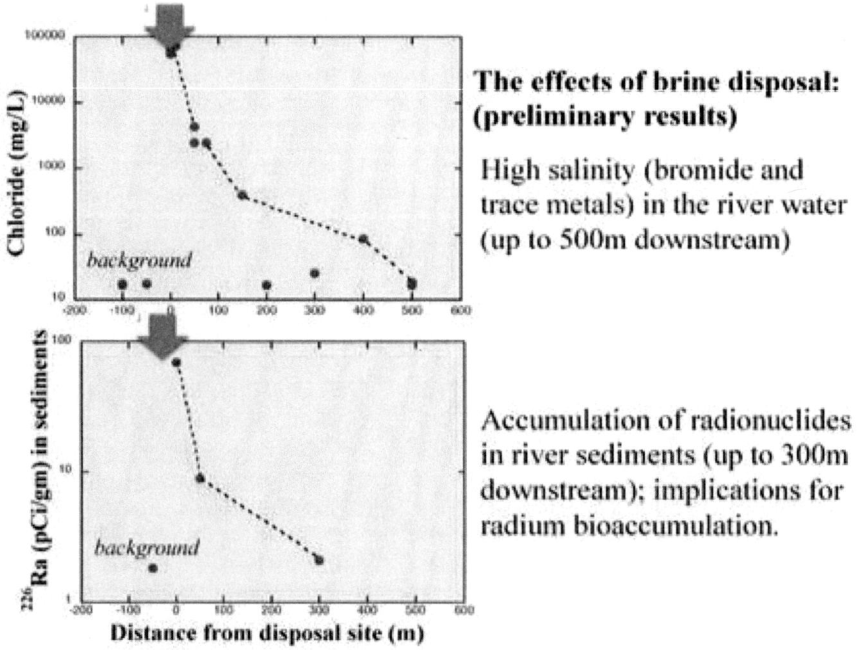

FIGURE 6-6 Downstream flow of chlorine from the outflow area of a treatment facility.
SOURCE: Jackson, 2012.

Another potential disposal strategy is to deliver produced water to a commercial wastewater treatment facility. It is a viable alternative, although these treatment plants do not always have the capacity to receive large volumes of produced water. Lack of familiarity with all the chemicals to be treated may be another limitation of this strategy.

Jackson described the practice of recycling or reusing produced water in a future fracturing job with or without treatment. This practice reduces the amount of water needed to be acquired and also reduces the expense of wastewater treatment. The challenge of reusing the water is monitoring the concentrations of chemicals in the produced water. Every time produced water is reused and collected, it can become more difficult to treat, and the contaminant levels can also be exceeded. Nonetheless, Dr. Jackson said that this practice is a positive development and the industry deserves credit for implementing it.

Quality of the Groundwater Naturally

Another question that Dr. Jackson attempted to answer is what is in shallow groundwater naturally? Jackson described a study that had as an objective to understand the quality of the groundwater in this region of the north of Pennsylvania. Figure 6-7 shows about 400 observations that have been collected over time in the area. The type of water was grouped into four types: two with low salinity levels and two with high salinity

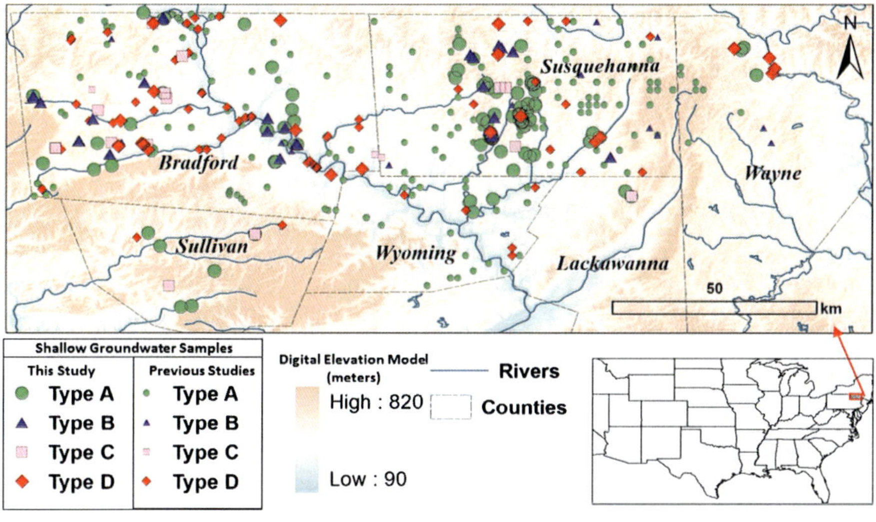

FIGURE 6-7 Occurrence of saline groundwater naturally enriched in barium and other elements in shallow aquifers.
SOURCE: Warner et al., 2012.

levels. More specifically, there is special interest in one type denominated Type D. It is one of the two high in salinity plus it has high bromine-to-chlorine ratio. This is important because those waters are the ones that look Marcellus-like and suggest natural connections to Marcellus-like brines through natural flow pass. It is not associated with drilling; it occurs naturally. The areas designated as Type D should be monitored for potential contamination.

Surface drinking water was also sampled and tested for dissolved gas concentration, salts, and NORMs. The results were published in Osborn et al. (2011) (see Figure 6-8). This is the first paper to look at the relationship between water quality and distance to gas wells. Most of the contaminants were not found. In the subset that was looked at initially, there was no evidence for the brines found naturally in deep formation waters or evidence for NORMs in residents' drinking water. What was found in some drinking water from wells were much higher dissolved gas concentrations of methane and ethane particularly. Within about 1 kilometer of the well, there is the likelihood of seeing very high concentrations of methane. Some wells fell within or above limits set by the Department of the Interior for hazard mitigation; immediate action is required on these wells because they pose a health hazard. In those cases, it is important to focus on the well integrity. According to historical records, 10 or 20 percent of the time, faulty cement or corrosion in the casing is present. Most cause of contamination or spillage is such a compromised well structure.

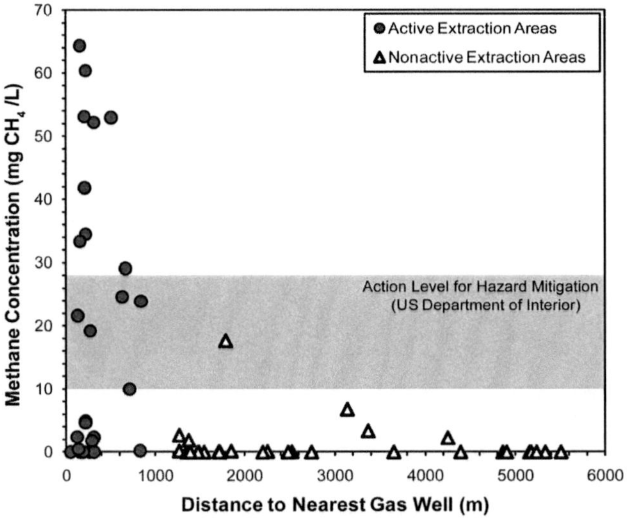

FIGURE 6-8 No evidence of brines or fracturing fluids was present, but methane concentrations in drinking water were higher near gas wells. The gray band is the Department of the Interior hazard mitigation recommendation.
SOURCE: Osborn et al., 2011.

The results of the study suggest that the liability distance be increased to 3,000 feet (see Figure 6-8). At that time, presumptive liability distance in Pennsylvania was 1,000 feet. Governor Corbett recently signed a bill (Pennsylvania Office of the Governor, 2012) stating that the new presumptive liability distance is 2,500 feet. The presumptive liability says that *if a homeowner has a problem with his or her water, now within a year of drilling, the operator is presumed guilty unless they can show otherwise.*

Other recommendations made in the study address disclosure of information to enhance environmental monitoring. For example, companies should release the isotopic values of the methane or ethane from each producing well (C-13 and the deuterium values), which could help rule out cases that could be perceived as contamination. Such testing would allow comparison of gas coming out of the ground and the gas that is in homeowners' wells; these data could help researchers identify sources of stray gas. Public disclosure and making information available in general would be a positive development. It could show transparency and willingness of industry to address public concerns.

As a last recommendation, the paper proposed studies on the health effects of chronic, low-level exposure to methane in people and animals (Jackson et al., 2011). There is little information in this area but it would be important for health care professionals to examine the issue.

Dr. Jackson acknowledged that while the session was focused on water, he wanted to briefly discuss natural gas and air. Dr. Jackson highlighted work done in collaboration with Nathan Philips at Boston University to map natural gas leaks across the city of Boston. The study found 3,300 gas leaks across the city (Phillips et al., 2013). A high-resolution methane imager is used to detect gas leaks, and samples are taken and analyzed in the laboratory for isotopic composition. The isotopic composition allows distinguishing among landfill gas, sewer gas, and natural gas coming from a pipeline. Figure 6-9 shows 3,300 gas leaks across the city. This type of information can help to reduce the environmental footprint of some of these processes.

Returning to the topic of water, Dr. Jackson briefly noted the situation in Pavillion, Wyoming. As mentioned earlier, one of the biggest public concerns is organics from fracturing fluids leaking into drinking water. Studies conducted by the EPA reported findings of dissolved gases and chemicals associated with hydraulic fracturing in the groundwater in Pavillion. Hydraulic fracturing there occurred as shallowly as about 1,000–1,500 feet underground; however, people may be obtaining their drinking water at 750 feet underground, sometimes in the same formation, this is not a good idea, Dr. Jackson said. Further test by the U.S. Geological Survey will confirm or refute contaminations.

In concluding his remarks, Dr. Jackson identified a number of positive developments on the water front:

- industry-driven initiative to recycle and reuse water for fracturing the next well;
- greater disclosure of the chemicals in fracturing fluids (except those that are trade secret), which is being driven by state laws; and
- interest in green completion and elimination of open wastewater pits.

These types of practices, if pushed, would benefit all, he said.

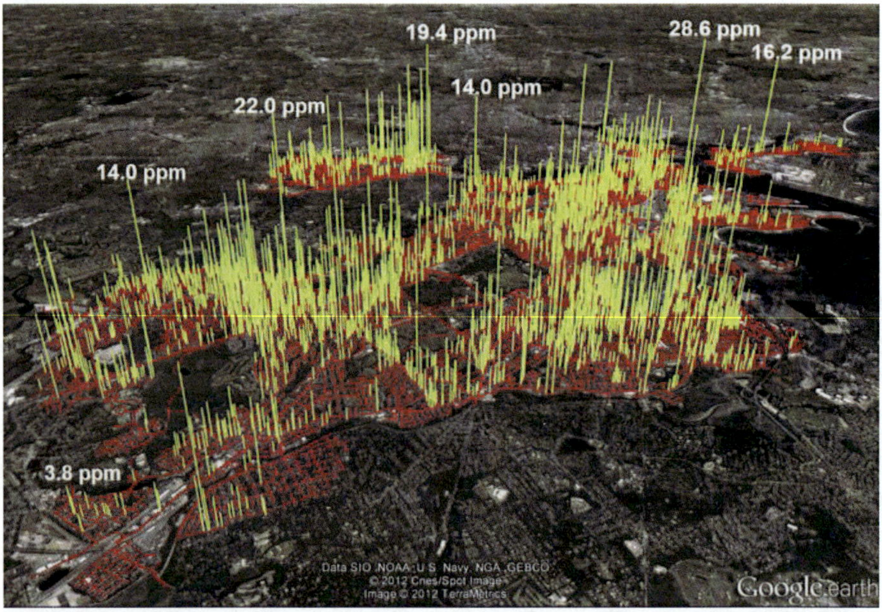

FIGURE 6-9 Emissions to the atmosphere: Methane leaks for the Boston metroplex.
NOTE: ppm = parts per million.
SOURCE: Phillips et al., 2013. Reprinted from *Environmental Pollution* with permission from Elsevier. Copyright © 2013.

EPA STUDY PLAN ON THE POTENTIAL IMPACTS OF HYDRAULIC FRACTURING ON DRINKING WATER RESOURCES: APPROACH TO STUDY POTENTIAL HEALTH IMPACTS

Jennifer Orme-Zavaleta, Ph.D.
Director, National Exposure Research Laboratory
Office of Research and Development
U.S. Environmental Protection Agency

Jennifer Orme-Zavaleta opened her remarks by saying that fracturing is not a new practice and the concept has been around for awhile. It has become more popular in the last decade because it has become part of the country's energy security and energy independence strategy. As the technologies evolve, the possibility of acquiring more of the shale gas deposits has become more viable. To clarify many of the concerns from the public and to develop a standardized practice for industry, Congress asked the EPA to study the impact of hydraulic fracturing on drinking water resources.[3]

Dr. Orme-Zavaleta explained that the objective of the study was specifically to assess whether hydraulic fracturing can affect drinking water resources and to better understand the factors that affect the severity and frequency of these impacts. The study focuses on surface and subsurface practices of hydraulic fracturing. The greatest attention was given to the well structure. Considering that most failures and accidents occur because of damaged or deteriorated well structure, this aspect needs to be deeply studied. She further discussed the research approach. There were five different research components to the study.

The first component is data gathering and analysis of available data. Data describe previous incidences of accidents at particular sites, their frequency, and how they were handled. The data gathered also include operating procedures from each one of the companies, the technology and types of materials they use, and the components of the fracturing fluid that they are using.

The second research component is based on case studies. The study identified several retrospective case studies as well as two prospective case studies. The purpose of looking back was to determine if drinking

---

[3] "*The conferees urge the Agency to carry out a study on the* relationship between hydraulic fracturing and drinking water, *using a credible approach* that relies on the best available science, *as well as* independent sources of information" (emphasis added). Department of the Interior, Environment, and Related Agencies Appropriations Act, 2010, H. Rep. 111-316. http://thomas.loc.gov/cgi-bin/cpquery/?&sid=cp111alJsu&r_n=hr316.111&dbname=cp111&&sel=TOC_351721& (accessed May 30, 2013).

water sources were previously affected and what factors were involved in those cases of contamination. The advantage of following the two prospective cases is that there is an opportunity to establish a prefracturing and predrilling baseline and to compare those baselines with the conditions afterward.

The third research component is failure scenario evaluation. This approach allows a comprehensive assessment and understanding of the impacts. It is important to look at issues such as water quantity, including water withdrawal, transportation, refracturing, and treatment and disposal of wastewater. She said that the EPA is not generally thought of as being interested in water quantity issues but she emphasized that it is not possible to look at water quality, an area that the EPA is known for, without considering quantity; these two characteristics are interrelated.

The fourth research component is laboratory studies. The evidence from laboratory studies can contribute to an understanding of the most efficient and safest wastewater treatment practices. Laboratory studies can be used to understand the interaction of hydraulic fracturing fluids and shale formations. There are different types of hydraulic fracturing fluids and each of the fluids interacts with the different types of shale formations. That these fluids are used with different components and at different concentrations must also be considered. Laboratory studies can also help assess the effectiveness of wastewater treatment. Laboratory studies can also help the treatment process. Processes must be able to effectively handle the types of contaminants that are in flowback and produced waters. If they are not effective, what would be the potential impact for drinking water resources? Other areas of concern are the analytical methods used and whether they are sufficient to measure the contaminants at the concentrations that are of interest. These issues are best determined and subject to experiments within the controlled laboratory environment.

The fifth research component is toxicity assessment. This component fundamental to understanding the interaction and impact of each of the chemicals used in hydraulic fracturing fluid (see Box 6-2). Toxicity assessments are focused on hydraulic fluids, wastewater, and naturally occurring substances that enter wastewater. The contaminants are being assessed for their chemical, physical, and toxicological properties. For some of the chemicals there is a lack of information; thus, an additional step of assessing the properties using quantitative structure–activity relationships or other computational types of approaches will be taken. This will help screen those chemicals and prioritize them for toxicity studies.

Studying the impact of hydraulic fracturing on drinking water is complicated, Orme-Zavaleta noted. To begin, the practice of hydraulic fracturing is not standardized. Every company is different. Each company's formulation of fluids is different. Different conditions require different types of fluids and mixtures, which makes comparisons complicated. Studies in this area must grapple with this issue.

BOX 6-2
Component Materials Used in Hydraulic Fracturing Fluids

Acids; Acid inhibitor
Biocides
Breakers
Buffers
Clay stabilizers
Corrosion inhibitors
Crosslinkers
Foaming agents
Friction reducers
Gelling agents
Iron control
pH-adjusting agents
Proppants
Scale inhibitors
Solvents
Surfactants

A complete list of chemicals as of November 2011 is available at http://www.epa.gov/hfstudy (accessed May 30, 2013).

Dr. Orme-Zavaleta also noted that the EPA does not have general regulatory authority over hydraulic fracturing fluids. Hydraulic fracturing is only regulated under the Safe Drinking Water Act if diesel fuel is used. When diesel fuel is used, a permit is required through the Underground Injection Control Program.[4] As was stated earlier, there is cause for concern if drinking water is contaminated with diesel because of human health effects. Diesel fuels contain benzene, toluene, ethylbenezenes, and xylenes, which are hazardous to health.

One of the study objectives was to determine specific indicators that would help the EPA track the chemicals in fluid and produced waters. These indicators can help determine standards for the industry such as frequency of use, toxicity of the chemicals, and improvement of the monitoring and detection methods.

In concluding her remarks, Dr. Orme-Zavaleta said that the biggest concern from the public and local authorities is the lack of understanding about whether hydraulic fracturing can impact drinking water sources and, consequently, human health. The objective of studies such as this one from the EPA, is to gather and analyze the available data, which can then be used to make informed decisions about the practice. It is known that many types of chemicals are mixed with water and subsequently injected in the ground. It is the responsibility of the EPA to know the toxicity and impact of the chemicals being used. Although the study will not conduct quantitative risk assessments, it will help understand the

---

[4] Water: Underground Injection Control, Regulation, 40 CFR Parts 144–148. Available: http://water.epa.gov/type/groundwater/uic/regulations.cfm (accessed May 30, 2013).

consequences of possible human exposure to the fracturing fluid, she said.

An important consideration of the study is to keep the public and industry informed of the processes under way. The information collected and available reports can be found through the EPA website.[5] The final report is due in 2014.

DISCUSSION

To begin the discussion, Bernard Goldstein asked Dr. Orme-Zavaleta whether the scenario component of the study she described would include an analysis of chemical mixtures. Dr. Orme-Zavaleta responded that the study was currently focused on individual compounds but that the case study component of the study will eventually allow for the examination of chemical mixtures. Christopher Portier asked Dr. Orme-Zavaleta if the study had preliminary results on endocrine disruptors in the chemicals identified in the study. Dr. Orme-Zavaleta responded that the data collection component of the study was currently identifying the chemical, physical, and toxicological properties of water quality components but was not yet at the stage of identifying endocrine disruptors. She highlighted that the health end points the study would focus on included carcinogenicity as well as developmental and reproductive end points, which would include, endocrine disruptors.

REFERENCES

Colborn, T., C. Kwiatkowski, K. Schultz, and M. Bachran. 2011. Natural gas operations from a public health perspective. *Human and Ecological Risk Assessment: An International Journal* 17:1039–1056.

EPA (U.S. Environmental Protection Agency). 2011. Plan to Study the Potential Impacts of Hydraulic Fracturing on Drinking Water Resources, EPA/600/R-11/122. http://water.epa.gov/type/groundwater/uic/class2/hydraulicfracturing/upload/hf_study_plan_110211_final_508.pdf (accessed May 30, 2013).

Jackson, R. B. 2012. *Hydraulic fracturing, water resources, and human health.* PowerPoint presentation at the Institute of Medicine workshop on the Health Impact Assessment of New Energy Sources: Shale Gas Extraction, Washington, DC.

[5] See http://www.epa.gov/hfstudy (accessed May 30, 2013).

Jackson, R. B., B. Rainey Pearson, S. G. Osborn, N. R. Warner, and A. Vengosh. 2011. *Research and policy recommendations for hydraulic fracturing and shale gas extractions*. Durham, NC: Center on Global Change, Duke University.

Moorman, R. 2010. Tamboran. Our commitment to the people of Ireland. http://www.tamboran.com/operations/ireland-uk/our-commitment (accessed May 30, 2013).

Nicot, J. P., and B. Scanlon. 2012. Time evolution of Barnett Shale well count and water use per well percentiles. *Environmental Science & Technology* 46:3580–3586.

Osborn, S. G., A. Vengosh, N. R. Warner, and R. B. Jackson. 2011. Methane contamination of drinking water accompanying, gas-well drilling and hydraulic fracturing. *Proceedings of the National Academy of Sciences of the United States of America* 108(20):8172–8176.

Pennsylvania Office of the Governor. 2012. Governor Corbett signs historic Marcellus Shale law. http://www.pikepa.org/Planning/Marcellus/Governor%20Corbett%20Signs%20Historic%20Marcellus%20Shale%20Law.pdf (accessed May 30, 2013).

Phillips, N. G., R. Ackley, E. R. Crosson, A. Down, L. R. Hutyra, M. Brondfield, J. D. Karr, K. Zhao, and R. B. Jackson. 2013. Mapping urban pipeline leaks: Methane leaks across Boston. *Environmental Pollution* 173:1–4.

Swackhamer, D. 2012. *Potential impacts of hydraulic fracturing on water resources*. PowerPoint presentation at the Institute of Medicine workshop on the Health Impact Assessment of New Energy Sources: Shale Gas Extraction, Washington, DC.

U.S. House of Representatives. 2011. Chemicals used in hydraulic fracturing. http://democrats.energycommerce.house.gov/sites/default/files/documents/Hydraulic-Fracturing-Chemicals-2011-4-18.pdf (accessed May 30, 2013).

Warner, N. R., R. B. Jackson, T. H. Darrah, S. G. Osborn, A. Down, K. Zhao, A. White, and A. Vengosh. 2012. Geochemical evidence for possible natural migration of Marcellus formation brine to shallow aquifers in Pennsylvania. *Proceedings of the National Academy of Sciences of the United States of America* 109(30):11961–11966.

7

Sustainable Energy

This chapter provides a summary of presentations that address the broad topic of sustainable energy. Both presentations emphasize the need to consider energy sources or technologies from a systems perspective. Each energy source can be considered within the broader context of the energy milieu which includes other available energy sources, the potential benefits and potential damages across the life cycle and into the future, and the community context. The presentations emphasize a systems perspective that encourages understanding the relationships among different fuels and strategies and identifying optimal sources for the given context.

SUSTAINABLE ENERGY FOR ALL?

Steven Hamburg, Ph.D., M.F.S.
Chief Scientist, Environmental Defense Fund

Steven Hamburg began his presentation by pointing out the need to consider hydraulic fracturing in a broader context. Hydraulic fracturing, he said, is an issue of energy, energy independence, environmental quality, health impacts, and, finally, the integration of these matters. The challenge is to not think in terms of one specific energy source or one specific technology. Losing the broader perspective by focusing too narrowly will result in erroneous science, ill health, and unwanted outcomes, he said.

Potential Energy Options

Dr. Hamburg suggested that there are many options for the energy future that can be considered. One option is nuclear power. There are both advocates for and opponents against building more nuclear power plants. The opponents list rational health- and safety-based arguments against nuclear power. However, the data on nuclear power from the last several decades, including recent examples of disasters, reveal that the rate of mortality and morbidity is lower for nuclear power plants than for coal-fired power plants. But that is not to say that there is not a small risk of something catastrophic occurring. A second option for energy is power derived from burning biomass. Similar to nuclear power, there are people who support using biomass for electricity, but there are plenty of people who believe burning biomass is a health disaster. A third option for energy is wind power. Again, some people support building onshore wind farms, coastal wind farms (e.g., Cape Wind), and, others fight the construction of these farms. Something similar is happening with hydraulic fracturing—some people support its use, and others oppose it. Dr. Hamburg posed two questions: Where should hydraulic fracturing fit among the energy mix utilized by the United States? And if advocates and adversaries of each energy source got together, what would be the outcome? Likely the status quo, he said, because there are legitimate issues with each one of these sources of power that have to be addressed. But they must be addressed in the context of the bigger picture—meeting our energy needs responsibly requires a suite of strategies in which negative impacts are minimized. If the energy future is focused too narrowly, it will not be successful, he said.

Every strategy mentioned, he noted, has a place in the energy future. This does not mean they are perfect, that they do not require good controls, or that more science is not needed. However, achieving a relevant balance through integration of these energy sources should be the objective.

Dr. Hamburg highlighted a state effort to reduce greenhouse gas emissions and consideration of a suite of potential strategies. California passed a climate change law, Assembly Bill 32 (California Environmental Protection Agency, Air Resources Board, 2013), which codifies the reduction of greenhouse gas emissions and calls for an energy future that will reduce emissions 80 percent below 1990 levels. Likely options to achieve this result include increasing energy efficiency, decarbonizing electricity generation, promoting smart growth, installing photovoltaic panels on rooftops, producing biofuels, electrifying vehicles and other entities currently not using electricity, and eliminating greenhouse gases from other sectors. He noted that there is a suite of approaches and they will likely need to proceed with all of them.

Shifting to climate change, Dr. Hamburg referred to work from Pacala and Socolow (2004) which divides the climate change problem into a series of different strategies. When mitigating climate change is the driver behind energy transitions, there is no specific solution or silver bullet. The approach is more akin to buckshot, and "all of the above" strategies to get reasonable outcomes. Certainly from a health standpoint, he said, it is important to curtail climate change because, going forward, climate change is likely to be one of the key drivers of negative health outcomes because of the disruption it will cause to social and biological systems.

Identifying Optimal Energy Sources

Dr. Hamburg suggested that moving forward, it is imperative to determine which energy sources are optimal in which situations. Achieving this requires more detail assessments. For example, electricity generated from coal combustion can be compared with electricity generated from natural gas combustion. Historically, only the emissions from smokestacks at the different power facilities were examined in this comparison. Now it is recognized that this is insufficient. Instead, how these fuels affect the environment and society throughout the supply chain must be considered. This includes the mining, production, and transport of coal and natural gas, as well as end use of the electricity generated, which for natural gas, includes use by industry and residential homes for heating.

He pointed out that it is necessary to understand the implications that a transition from coal to natural gas will have on climate change. Methane is the main constituent of natural gas, and there is methane leakage at each point along the natural gas supply chain—production, processing, transport, and combustion. Methane is also a greenhouse gas, and its leakage should be accounted for in the comparison of natural gas and coal. The challenge is how to make methane leakage equivalent to carbon dioxide emissions—which persist in the atmosphere for very different lengths of time—for a true comparison. It is possible to use the 100-year global warming potential (GWP) of methane,[1] which is 21[2] (EPA, 2013b), but GWP assumes that the short term is not relevant because the outcome is assessed 100 years from now. More specifically, GWP is the impact of a single pulse emission 100 years after it is released. A useful analogy for GWP is worrying about the impacts of renting a car today 100 years in the future. What are the implications 100 years from now of a pulse of emissions from a power plant? But a hundred years from now is an abstraction. There is a lot of time between now and then, and most people care about what happens between now and 100 years from now. What we really care about are the implications of owning a particular type of car for its lifetime, or even more important what are the implications of changing the characteristics of an entire fleet of cars (e.g., Corporate Average Fuel Economy [CAFE] standards). GWP, as traditionally applied, does not reveal much about the impacts of a power plant over time. Most people care about what the power plant does over its functional life or the effects of its emissions over the next 20 years, as well as its impacts over the longer term.

---

[1] Global warming potential was developed to compare the ability of each greenhouse gas to trap heat in the atmosphere relative to carbon dioxide.

[2] It should be noted that the Intergovernmental Panel on Climate Change uses 25 for the GWP (100-year) of methane; for a series of reasons, the U.S. Environmental Protection Agency uses an outdated factor.

Dr. Hamburg explained that when a comparison of the effects of a pulse emission from a coal power plant and a pulse emission from a natural gas power plant is performed, it is revealed that there is an immediate 20 percent reduction in net radiative forcing (the change in the balance between radiation coming into the atmosphere and radiation going out) by switching from coal to natural gas (Alvarez et al., 2012). Over 200 years, the benefit is a 45 percent reduction in net radiative forcing for the natural gas fueled electrical power plant in comparison to the one using coal. The climate impacts of fuel switching are best understood by considering their implications continuously over time for a large-scale shift in fuel. The concept of Technology Warming Potential introduced by Alvarez and colleagues (2012) allows one to make these comparisons simply. Using this more robust framework, Hamburg stated that there is a climate advantage from switching to natural gas from coal immediately and over time for the electricity sector. Yet it is important to note that the lower the methane leak rate across the supply chain the larger the climate benefits of such a shift, so long as the leak rates are below about 3 percent.

It should be noted, Dr. Hamburg said, that this calculation assumes methane leak rates estimated from the U.S. Environmental Protection Agency (EPA) data based on a natural gas study done nearly 20 years ago and updated with more recent activity factors (EPA, 2013a); even though industry techniques have changed radically since then, these are currently the best data available. Researchers and industry are working to complete a series of studies to collect empirical field data in order to populate these calculations with empirical numbers, rather than best estimates, allowing for a better assessment of the health impacts of these comparisons.

An interesting topic of political relevance, Hamburg suggested, is the comparison of natural gas versus gasoline for transport and its climate implications. When natural gas is compared with gasoline for the conversion of a fleet of cars, there is an immediate 30 percent disadvantage based on currently available data (e.g., EPA estimated leak rates and engine efficiency). It would take about 85 years before there is a climate change benefit to switching from gasoline to natural gas for a fleet of vehicles. He noted that a comparison of natural gas and diesel can also be performed. Because diesel holds more energy per unit of carbon dioxide emitted, the advantages of switching to natural gas are not as great. It would take more than 200 years to see any climate benefits assuming 2010 EPA estimated methane leak rates and literature estimates of engine efficiency (Alvarez et al., 2012).

Dr. Hamburg also suggested that from a policy standpoint, it would be helpful to know what leak rate is required to make these conversions beneficial relative to climate change. It is possible to compare well-to-wheels leak rates and the number of years that must elapse before a climate change benefit is realized. If the well-to-leak rate is 1 percent, a benefit is realized immediately. Analytically, it is now possible to solve for when the benefit will be accrued, which allows society to decide which fuel is preferable. He stated that there may be a short-term disbenefit and a long-term benefit. But why not strive for a leak rate that makes benefits available continuously, especially if a significant capital investment is going to be made? Minimizing methane leaks would have health and climate benefits, and would represent a victory for society and for industry (because they collect more product and potentially reduce operating and maintenance costs through a better understanding of sources of leaks).

Dr. Hamburg explained that the analytical work described here is necessary to understand the relationships among different fuels and strategies. This work will allow a clearer understanding of what will meet social and legal goals and what will not, and when they affect health. This is only possible when all types of fuels are analyzed across the landscape in a comparative way—requiring one to understand the life-cycle implications of diverse fuels in greater detail than currently available.

Including Biomass in the Energy Future

Another energy source that is hotly contested because of its potential impacts on health is bioenergy, Dr. Hamburg said. A forest may be considered a pristine place, but it might have been a pasture in the previous 100 years. It is important to think of many of the forests of the United States as cultural landscapes—a landscape heavily influenced by people. The way a landscape is used can be affected by the political economy and can have social effects. The concept of the cultural landscape allows us to understand that often there is much less of a perceived conflict between nature and people than might otherwise be the case. It also allows one to understand how people have used the land and how they might use it in the future with or without negatively impacting the environment.

Biomass energy plants can convert municipal forest waste from urban environments to energy, burning material that might otherwise simply decay, and use it for heat in a highly efficient manner. There are health effects of particulate emissions, but if larger biomass plants implement effective particulate controls there is an opportunity for a win-win: low carbon energy with limited health and environmental impacts. Understanding if such plants can actually be deployed represents the kinds of issues and trade-offs that need to be considered.

Dr. Hamburg emphasized that the bottom line in developing a low-carbon economy is not about deploying a single strategy. It is not about a single fuel. It is about optimization among them all, deploying a mixed fuel strategy is more complicated and thus more difficult, but has the potential to more effectively meet our climate goals while reducing health impacts. It is important to understand the context of each fuel, its implications, when it is appropriate to deploy, when it is better than another fuel, and when it is worse. It is about "all of the above."

SUSTAINABLE ENERGY FOR ALL: ENSURING HEALTH THROUGHOUT THE ENERGY PRODUCTION AND USE LIFE CYCLE

Daniel S. Greenbaum, M.S.
President, Health Effects Institute

Daniel S. Greenbaum began his presentation by reiterating Dr. Hamburg's point that discussions about shale gas extraction need to be considered in the larger context of a comprehensive energy discussion where comparisons are made across all energy forms. America's production and use of energy result from a complex system of supply and demand. It also creates a complex web of potential health, environmental, and other effects throughout the life cycle. Any one component of the system (e.g., shale gas hydraulic fracturing) must be placed in the context of the whole system. This approach requires that effects be evaluated throughout the system, and the effects be compared on an "apples-to-apples" basis across different energy sources and uses.

One example of a broad, systems approach to energy analysis is the National Research Council (NRC) report, *Hidden Costs of Energy: Unpriced Consequences of Energy Production and Use* (NRC, 2010). Congress requested this study as part of the Energy Policy Act of 2005 (Public Law 109-58). The task of the report was to "evaluate key external costs and benefits—related to health, environment, security, and infrastructure—that are associated with the production, distribution, and use of energy but not reflected in the market price or fully addressed by current government policy." That is the essence of what economists consider an "externality"—an effect that is not paid for, but that has a cost.

The report concluded that, in the United States, there are many externalities related to energy production and use. The committee that authored the NRC report assigned monetary values to a wide range of damages, although an equal number or perhaps greater number of external effects could not be monetized. This approach allowed for some degree of "apples-to-apples" comparison between different forms of energy. The overall monetized damages in 2005 were $120 billion, but that number does not incorporate damages due to climate change, Mr. Greenbaum noted.

To assess the monetary values of damages, the report focused on several key components of the energy system. These areas included electricity generation, transportation, and heating for buildings and industrial processes. These three areas combined account for approximately 80 percent of the energy use in the United States. The report also described sets of external costs for infrastructure and national security that are not always embedded in the market price. Whenever possible, the report examined the full life cycle of the energy source and external costs. To provide a longer-term view, the report looked at both actual damages in 2005 and projected damages in 2030.

For the nonclimate damages, a fairly conventional method of looking primarily at air pollution was used, because those data were most readily available. Emission levels and ambient concentrations of air pollutants could be estimated. Exposures of people to these pollutants and the effects of these exposures could also be calculated. These evaluations permitted the assignment of monetary damage values to these effects. Modeling was used to estimate damages based primarily on sulfur dioxide (SO_2), nitrogen oxides (NO_x), and particulate matter (PM) emissions across the 48 contiguous states. The effects that were examined included damages to human health; grain, crop, and timber yields; building materials; recreation; and visibility of outdoor vistas. The single largest contributor to the damage estimates was related to human mortality, despite the fact that the highest value for a statistical human life (recently used by the EPA and others) was not used in this assessment (NRC, 2010).

Damages from Electricity Generation

Mr. Greenbaum discussed the monetary value of damage associated with energy sources and by sectors. In 2005, coal used for electricity generation accounted for $62 billion in nonclimate damages. Of 406 power plants, 10 percent (the oldest, largest—which produced 25 percent of net generation) were responsible for 43 percent of the damages. This variation in damages is primarily due to a disparity in the tonnage of emitted pollutants. The total amount of nonclimate damages equates to an external cost of 3.2 cents per kilowatt-hour, which is not an insignificant addition to the cost of electricity. For the analysis carried out to 2030, it is assumed that existing rules will be successful in reducing emissions of traditional air pollutants: emissions of SO_2 and NO_x per kilowatt-hour are expected to fall by 64 percent and 50 percent, respectively. The 2030 external cost decreases to 1.7 cents per kilowatt-hour, despite rising incomes and an increase in the value of a human life. Thus, it is possible to internalize those external costs and reduce them.

A very different outcome was observed with natural gas. For the 498 natural gas-fired plants, which account for approximately 71 percent of domestic natural gas generation, there were $740 million in nonclimate damages. This is slightly more than 1 percent of the damages associated with coal. This finding is due to the much lower levels of pollutants emitted during natural gas combustion. The 10 percent of power plants with the greatest damages were older plants; these accounted for 65 percent of the damages. The cost per kilowatt-hour for natural gas is one-twentieth that associated with coal, or 0.16 cent. By 2030, when newer, cleaner plants come online, the cost will decrease to 0.11 cent per kilowatt-hour. This decrease is due to an expected 19 percent reduction in NO_x emissions and 32 percent reduction in PM emissions per kilowatt-hour (NRC, 2010).

It is beneficial to put these data in a different perspective and investigate where the damages are localized, which is also helpful for health impact assessments. The majority of coal-fired power plants are in the eastern United States. In contrast, natural gas plants (and their concomitant damages) are spread more widely across the country, Mr. Greenbaum said.

Damages from Transportation

When examining the transportation sector, the committee focused on highway vehicles, which account for 75 percent of energy use within this sector. Various fuels were considered: oil (both petroleum gasoline and diesel), natural gas, biomass or biofuels, and electricity. A full wells-to-wheels analysis was performed that incorporated the extraction of the feedstock, the transport of the feedstock to the refinery, the fuel conversion and refining process, the transport of the fuel to the pump, the manufacturing of the vehicle (which is often not included), and the tailpipe and evaporative emissions from operating the vehicle.

It was determined that the aggregate nonclimate damages in 2005 from transportation were $56 billion. Light-duty vehicles accounted for 60 percent of these damages. Per gallon, damages were estimated to be 23 to 38 cents per gallon, which is 1.1 to 1.7 cents per vehicle-mile traveled. This cost may not seem high when the price of a gallon of gasoline is $3.80, but the costs accumulate when more than 3 billion barrels of gasoline are used for transportation annually (EIA, 2012).

Mr. Greenbaum highlighted the finding of minimal variation across the different technologies and fuels analyzed. Some (electricity and corn ethanol) had marginally higher levels of damages whereas others (cellulose and natural gas) had slightly lower life-cycle damages. This finding should be interpreted cautiously: the damages associated with electric cars are mainly due to the coal-fired power that supplies electricity in much of the country and is associated with considerably higher damages. It is also noteworthy that damages were not spread evenly among the different life-cycle phases. In most cases, vehicle operation accounted for less than one-third of the total damage. Vehicle manufacturing was a significant contributor to damages.

Looking forward to 2030, the minimal variation among fuels and technologies will shrink even further. This shift is due to new fuel economy standards that will increase vehicle efficiency, diesel emission rules that will reduce NO_x and PM levels, and electricity-generating power plants that will become cleaner and more efficient (as discussed previously).

Damages from Greenhouse Gas Emissions

The committee that authored the NRC report also estimated climate-related emissions for the sectors described above. However, a specific damage estimate was not ascribed for climate damages. The committee instead reviewed a range of analyses in the climate-change literature that have used integrative assessment models to try to assess the social cost of carbon. One of the limitations in estimating the value of specific damages, and a reason why the committee did not pursue this further, came from the wide range of values assigned to the cost per ton of carbon dioxide equivalent (ranging from $1 to $100). The committee found that the key factors responsible for this variation were (a) the rate at which future damages are discounted and (b) how fast damages (as a percentage of gross domestic product) were predicted to increase with temperature.

Within the electricity sector, natural gas produces half the carbon dioxide emissions of coal—coal emits 1 ton of carbon dioxide per megawatt-hour of power generated, and natural gas emits 0.5 ton of carbon dioxide per megawatt-hour. This is not anywhere near the discrepancy observed for the nonclimate change damages, where damages were 20 times greater for coal compared with natural gas. Nuclear, wind, solar, and biomass sources were also investigated. Life-cycle emissions of greenhouse gases from these energy sources were so small as to be negligible compared with those from fossil fuel-generated electricity.

For transportation vehicles, there was no major variation across the technologies in terms of greenhouse gas emissions. Some benefits were observed for cellulosic ethanol, but tar sands petroleum and Fischer-Tropsch diesel emitted more carbon dioxide per vehicle-mile traveled. Vehicle operation, in most cases, is a substantial contributor to the total life cycle of greenhouse gas emissions. The projections for 2030 show even closer estimates in the greenhouse gas emissions per vehicle-mile traveled between fuels and technologies; this is due to substantial improvements in fuel efficiency.

As mentioned above, the damages per ton of carbon dioxide-equivalent ranged from $1 to $100 and the committee did not estimate climate damages. However, a few arrays are presented for a point of reference (see Table 7-1). In selecting $30 per ton—a moderate estimate of climate damages—and combining this with the nonclimate damages, the impact of coal-fired electricity generation nearly doubled to approximately 6 cents per kilowatt-hour (compared with 3.2 cents when

TABLE 7-1 Combining Nonclimate and Climate Change Damage Estimates (2005)

Energy-Related Activity (fuel type)	Nonclimate Damages	Climate Damages (per ton CO_2 equivalent)		
		@ $10	@ $30	@ $100
Electricity Generation (coal)	3.2 cents/kWh	1 cent/kWh	3 cents/kWh	10 cents/kWh
Electricity Generation (natural gas)	0.16 cent/kWh	0.5 cent/kWh	1.5 cents/kWh	5 cents/kWh
Transportation	1.2 to ~1.7 cents/VMT	0.15 to ~0.65 cent/VMT	0.45 to ~2 cents/VMT	1.5 to ~6 cents/VMT
Heat production (natural gas)	11 cents/MCF	7 cents/MCF	70 cents/MCF	700 cents/MCF

NOTE: kWh = kilowatt-hour, MCF = thousand cubic feet, VMT = vehicle miles traveled.
SOURCE: NRC, 2010.

nonclimate damages were considered alone). For electricity generated by natural gas, climate damages (based on $30 per ton of carbon dioxide equivalent) resulted in an external cost of 1.5 cents per kilowatt-hour, which, given that the nonclimate damages were so low originally, is a vast increase in damages. The transportation sector also experienced an increase in the damage estimate, to approximately 1.6–3.7 cents per kilowatt-hour, when climate impacts were considered.

Evaluating Energy with a Systems Approach

The *Hidden Costs of Energy* report found that nonclimate damages from electricity generation and transportation exceeded $120 billion in 2005. These damages were principally related to emissions of SO_2, NO_x, and PM. The committee believed that the total value was a substantial underestimate because it did not include damages related to climate change effects, the health effects of hazardous pollutants, ecosystem effects, or the external effects on infrastructure and national security.

Economists assert that estimating a cost does not imply that the cost needs to go to zero. It is important to consider the marginal costs—the cost of diminishing a burden compared with the value added from the reduction. For instance, if it costs $100,000 to get the next $1,000 of reduction, this might not be the best option for society as a whole. Still, there was evidence from these analyses that showed decreasing emissions, improving energy efficiency, and shifting to cleaner methods of generating electricity could reduce damages and have a substantial benefit.

Returning to the topic of hydraulic fracturing, it is important to place natural gas in the larger—and life-cycle—context. In the analyses found in the NRC report, natural gas was a favorable option in the nonclimate area and even somewhat beneficial in the climate area; however, there are still significant data challenges and questions that remain. For example, much is still unknown about the upstream effects of natural gas and coal, and it is not yet possible to quantify many environmental effects (such as water requirements for biofuel production) on a national scale. The bottom line is that energy cannot be handled with a "one solution by one solution" approach. Systems approaches to these energy questions—like the one outlined here and the "all of the above" approach presented by Dr. Hamburg—are needed to make fully informed decisions in the future.

DISCUSSION

Following the presentations, Lynn Goldman began the discussion by asking the presenters to comment on energy renewability, energy security, and incentives. That is, the energy sources that are renewable and also can be produced domestically and are the focus of large tax incentives to encourage them. The presenters were also asked to comment on biomass combustion, which has also been incentivized by tax policy and encourages the growing and burning of trees. Dr. Hamburg responded that there is feedstock (biological material that can be used directly as fuel) that is available and that has minimal impacts on the climate and that would be beneficial to the economy of the forest. Hamburg pointed out that the wood pulp industry, for example, is hurting because the public is reading fewer newspapers and this is causing problems for the low-grade wood market. If low-grade wood is not going to pulp for newspapers, then bioenergy is a great use for it; you can produce bioenergy. It is a matter of having the right rules and the incentives aligned with the rules so that the forest of today is not turned into the forest of the past century. In the past, there was a wave of cutting down almost every tree from East Coast to West Coast. That could be repeated with bad incentives. He stated that this does not mean that there should be no incentives or no use of forest material, but that incentives should be considered for their potential to create perverse outcomes.

John Balbus asked the presenters how to use systems thinking to produce policies that are not just economically optimized but that balance trade-offs (produce energy but not at the expense of environmental justice, for example). Dr. Hamburg responded that a criterion to address local impacts is needed. For example, a local heating and distribution biomass plant is located in the middle of downtown St. Paul, Minnesota. It provides cheap, local heating but must abide by their air pollution permit. Without the right permits, it would increase pollution and morbidity in the local area. The challenge is finding the sweet spot of balance between the two—pollution controls that are not so high that they produce barriers to siting the plant in the community and the protection of the local community.

Richard Jackson noted that the use of gas for heat in homes is very high, yet there is not a focus on producing high-quality gas for homes. He asked the presenters to comment on the lack of progress in this area. Dr. Hamburg responded that Dr. Jackson's question implied that the gas used in homes is a low-grade end product, but the real issue is that high-efficiency furnaces are needed. He commented that if you have a boiler or furnace that is at 97 percent and it provides direct heat and there are good controls so that it is used wisely—that is optimal thermodynamically. Investment in efficient boilers and furnaces, he noted, is far more efficient (given that heat is not lost through window leaks) than burning electricity and bringing electricity into the home.

Luiz Galvão asked the presenters to think about key policy recommendations they felt would be impactful. Both presenters emphasized that they are not in a position to make policy recommendations and they do not speak for their agencies. That said, Mr. Greenbaum said that his focus would be on greening the electricity system. He noted that there is a problem as long as there are coal-fired plants and they are not being replaced by renewable and other sources of energy. There are immediate health issues associated with coal versus other sources of electricity; further, the relative benefit of electric vehicles is undermined because the electricity these vehicles run on is generated by coal. The life-cycle damage from those vehicles is affected if energy is ultimately generated by coal. Dr. Hamburg offered the development of technology-neutral incentives and a set of pollution filters and other criteria to test new technologies (e.g., thermodynamic system implications for health and other local impacts) as his policy suggestion.

Henry Anderson asked the presenters if it was better to encourage the use of plentiful biomass for home heating, especially in less populated areas with available wood resources or move to gas. Shifting to gas would require adding gas lines to low-density areas and would have associated economic costs. On the other hand, continuing to use a wood fire boiler, for example, is not as energy efficient and contributes to pollution. Dr. Hamburg responded that, although not an expert in this area, he believed that for larger institutional settings (such as schools and universities) heating with biomass (with good pollution control on institutionally based boilers) would be a net winner financially and environmentally for the communities described. As long as the waste is harvested, it would have a minimal net impact on the forest and still provide jobs and money.

Nicholas Jones, a public health physician from New Zealand, told the panel that a large oil plate had been discovered in his area. He was attending the workshop to learn about the potential health issues associated with natural gas. He told the panel that in New Zealand the electricity supply is based on about 60–70 percent renewable energy. He asked whether the benefit of a shift to natural gas would be less given this context. Mr. Greenbaum responded that the single largest advantage of natural gas in the U.S. context has been in comparison to coal as a source of electricity because of its contribution to air pollution and other factors. He stressed that Dr. Jones was asking the right question—How would the energy source work within the given context? He noted that even within the U.S. context, there are few coal-fired plants in the West, so the benefit from shifting to natural gas would not be as beneficial as in the East, which has many more coal-fired plants.

REFERENCES

Alvarez, R. A., S. W. Pacala, J. J. Winebraker, W. L. Chameides, and S. P. Hamburg. 2012. Greater focus needed on methane leakage from natural gas infrastructure. *Proceedings of the National Academy of Sciences of the United States of America*. doi:10.1073/pnas.1202407109.

California Environmental Protection Agency, Air Resources Board. 2013. Assembly Bill 32: Global Warming Solutions Act. http://www.arb.ca.gov/cc/ab32/ab32.htm (accessed May 30, 2013).

EIA (U.S. Energy Information Administration). 2012. Frequently asked questions: How many gallons does the U.S. consume? http://www.eia.gov/tools/faqs/faq.cfm?id=23&t=10 (accessed May 30, 2013).

EPA (U.S. Environmental Protection Agency). 2013a. National greenhouse gas emissions data. http://www.epa.gov/climatechange/ghgemissions/usinventoryreport.html (accessed May 30, 2013).

EPA. 2013b. Overview of greenhouse emissions: Methane. http://epa.gov/climatechange/ghgemissions/gases/ch4.html (accessed May 30, 2013).

NRC (National Research Council). 2010. *Hidden costs of energy: Unpriced consequences of energy production and use*. Washington, DC: The National Academies Press.

Pacala, S., and R. Socolow. 2004. Stabilization wedges: Solving the climate problem for the next 50 years with current technologies. *Science* 305(5686):968–972.

8

Research Opportunities: Research Community

During the workshop, a panel of scientists was convened to discuss the research opportunities and needs for understanding health impacts of hydraulic fracturing. The presenters were asked to reflect on the evidence presented the previous day and identify the next steps to minimize health effects particularly as technology evolves. Further, the panel was asked to identify in their opinions where there is uncertainty in the available evidence. The purpose of the panel was to elicit a range of viewpoints, but not to reach a consensus. The following section is a summary of the initial presentations and the discussion among the panelists.

SUMMARY OF PRESENTATIONS

David Carey, Ph.D.
Director, Weis Center for Research
Geisinger Health Center

David Carey provided background on the Geisinger Health System, located in Danville, Pennsylvania—part of an area affected by hydraulic fracturing. He described that leadership and staff have a vested interest in the health outcomes occurring regionally related to the use of hydraulic fracturing. In response to increased activity, the health system assembled a coalition of stakeholders and experts from academia, other health care systems, government, and industry to conduct a coordinated, multi-disciplinary research project. Carey stated that the focus of the project is to collect data to inform a scientifically rigorous assessment of regional health impacts. At one level Geisinger is creating a database of health information on patients in the region that could be used for broad surveillance and analysis to identify areas of concern and "hotspots" in terms of adverse health outcomes. These data could then be used as the basis for performing more focused hypothesis-driven studies. Combining this research with geographic information systems mapping, patient and clinical data can be located in both space and time.

Geisinger has a number of key system attributes that position it to be ideal to play a central role in this effort, including being a highly integrated health care system and providing patient care to a large number of patients covering 31 counties, including areas where intense drilling activity is occurring. The integrated system comprises 4 hospitals and 40 community-based clinics that provide both primary care and specialty care, and Geisinger also operates as a health care insurance company. This degree of integration allows for cross-disciplinary research along the health care front and to better serve the health needs of the patients in the region. Further, Geisinger was one of the first adopters of electronic health records for both inpatient and outpatient data in the country. The system has comprehensive data on hundreds of thousands of patients beginning in 2004. The data are stored in a clinical data warehouse, a more easily searchable and mineable platform.

Additionally, the Geisinger Health System has become a leader in the Keystone Health Information Exchange and the Beacon Community Project which electronically links health care providers in the region. Dr. Carey said that Geisinger is initiating an effort to expand this collaboration further to engage other health care systems and other providers in the region to participate through this health information exchange network. With the ability to collect larger sets of data, standardize the data, and blend them into the searchable database, longitudinal data mining could be possible. Understanding the health impacts is a complex problem. Dr. Carey described that although the focus of the presentation at this workshop is on health impacts, he believes that there is also a need to incorporate information on environmental assessment research.

Rob Donnelly, M.B.Ch.B., MFOM
Vice President of Health
Royal Dutch Shell

Royal Dutch Shell is engaged in discussions such as this workshop, according to Rob Donnelly, in order to be at the table and participate in learning of potential health impacts related to its work. Dr. Donnelly provided an overview of the approaches Shell is using to engage in discussions with Shell communities, who they recognize have real concerns.

Dr. Donnelly noted that as stated earlier, natural gas is a part of the energy mix for the future. Royal Dutch Shell's focus, therefore, must be on conducting its work right. The company has issued a set of operating principles for tight gas production where hydraulic fracturing is used. The operating principles were shared publicly in 2011 and continue to be adopted across Shell's global onshore operations. The principle focus is on safety and well integrity, water, air, footprint, and community. Dr. Donnelly stated that although there is a need for additional research, these focus areas highlight what can be done currently to protect human health based on available information while working to ensure growing energy needs.

Incidents of the last 2 years have highlighted the primacy of well integrity for any hydrocarbon operation according to Dr. Donnelly. Each of Royal Dutch Shell's wells is individually designed and pressure tested before being put into production. He went on further to state that the company focuses on water recycling and protection. Dr. Donnelly emphasized that Royal Dutch Shell works with local communities, to understand water supply issues in the area and to ensure that corporate decisions take into account the entire water resource in an area. For air quality and reducing air emissions, he described that the company's effort to add fitted catalytic technology to diesel generators in some areas. According to Dr. Donnelly, the company establishes an interactive dialogue early on with the community to discuss local concerns, for example, rerouting delivery trucks to limit traffic around schools or through the middle of town. The company's community principle utilizes health impact assessment to identify socioeconomic impacts. Incorporating these simple things in addition to research studies is a key approach to obtaining gas safely, according to Dr. Donnelly.

Bernard D. Goldstein, M.D.
Professor Emeritus, Department of Environmental and Occupational Health
Graduate School of Public Health, University of Pittsburgh

Bernard Goldstein began by describing, in his opinion, the unfortunately typical progression of environmental and occupational issues related to human health. A new technology is not adequately evaluated for potential adverse health consequences. Public concern follows that results in calls for investigation of a potential causal relationship between the new technology and adverse health consequences, which is then usually hampered by inadequate exposure and toxicity information to perform a retrospective analysis. Dr. Goldstein identified barriers: many changes occur rapidly over time, disease clusters occur whether causal or not, litigation occurs, and message control can be harmful in terms of getting research efforts started. The end result is usually that solutions are delayed. Over time, industry will find technologies to reduce pollutants, which is ultimately in their best interest.

Specific to hydraulic fracturing, Dr. Goldstein agrees with earlier statements that the public is confused and concerned. For instance, is hydraulic fracturing old or new technology and does it cause groundwater contamination? Some statements from the government and industry might suggest that it is a new technology that now allows extraction of natural gas. However, in response to questions about health and safety issues, the message from the same sources is not to worry because the technology has been around for more than 40 years. It cannot be both, Dr. Goldstein remarked. Second, the public receives conflicting information about whether hydraulic fracturing causes groundwater contamination. The focus too often is on whether the successful release of hydraulic fracturing agents 5,000 feet underground will cause groundwater contamination, which is not responsive to the public's question of whether water pollution will occur over the entire process from the development phase through 30 years from now when the well is no longer active. Dr. Goldstein suggested that the process likely will not affect surface water if chemicals are released 5,000 feet underground, but there is a possibility of chemicals seeping into groundwater when casings blow, drums leak, and trucks spill. Dr. Goldstein noted that three advisory committees, established in 2011 by President Obama and the governors of Maryland and Pennsylvania to examine the issue of unconventional shale gas drilling, have in their executive orders requests for advice on the protection of public health. But, of the 52 members appointed to the three commissions, not one has a health background. Not surprisingly, very little has come out of their advice regarding health research. For example, in Pennsylvania, a bill passed after advice from this commission includes 17 different state agencies that will receive funding allocations from the Pennsylvania impact fee, but not one penny went to the Pennsylvania Department of Health to develop standard public health surveillance or to fund research. Similarly, in President Obama's Advisory Commission, the lead is given to the Department of Energy with input from the U.S. Environmental Protection Agency and from the Department of the Interior. The Department of Health and Human Services is not included (Goldstein et al., 2012).

Finally, Dr. Goldstein expressed that the disclosure of the chemicals used in hydraulic fracturing has not been transparent. There has been a significant increase in information released about the hydraulic fracturing chemicals with the adoption of the recent laws. However, there are still inappropriate exceptions for confidential business information. Of greater toxicological concern than the hydraulic fracturing chemicals are the chemicals brought up from underground that need to be disposed. These include naturally occurring brine constituents, radionuclides, and other potentially toxic agents. In Pennsylvania, the exceptions to releasing chemical information are stated as follows:

"Notwithstanding any other provision of this chapter, a vendor, service provider, or operator shall not be required to do any of the following:

1. Disclose chemicals that are not disclosed to it by the manufacturer, vendor, or service provider.

2. Disclose chemicals that were not intentionally added to the stimulation fluid.
3. Disclose chemicals that occur incidentally or are otherwise unintentionally present in trace amounts, may be the incidental result of a chemical reaction or chemical process or may be constituents of naturally occurring materials that become part of a stimulation fluid."[1]

These exceptions result in health scientists not having access to necessary information. In summary, the issues that Dr. Goldstein described as being of most concern to toxicologists are the agents used in hydraulic fracturing; natural gas hydrocarbons, the naturally present agents brought to the surface in flowback water (e.g., arsenic, brine components, radionuclides), and the effects of mixtures of these agents and reactants.

Roxana Witter, M.D., M.S.P.H.
Assistant Research Professor,
Environmental and Occupational Health
Colorado School of Public Health

Roxana Witter described a number of areas where community impact research could be done. There are preliminary reports that link community impacts to natural gas developments, but the effects on public health are uncertain. Sociological and socioepidemiological literature demonstrates that social environment can affect health and this literature may suggest appropriate methodologies for future research. She added that there is some initial evidence for social impacts as a result of population influx, including traffic, noise, rise in sexually transmitted infections, changes in economic conditions, and quality-of-life effects that will likely affect individual and community health. Dr. Witter suggested that looking into the longitudinal impacts pre- and post-extraction by looking at similar metrics, such as sexually transmitted infections, crime, substance abuse, and so forth, and looking at impacts from population influx would be revealing. Further, surveys could be used to understand social cohesion and social capital impacts at a community level and to direct community interventions. Another method she suggested could utilize the stress end points and identify potential markers, such as heart rate, cortisol, and C-reactive protein. The purpose of these surveys could be to assess perceived stress (measuring acute stress) and affect (measuring chronic stress). Further, surveys could be used to look at decreased exercise, increased substance abuse, and so forth to understand the exposures that cause stress. Finally, Dr. Witter suggested that a variety of individual

---

[1] Pennsylvania Consolidated Statutes, Title 58, Oil and Gas, § 3222.1(c), Disclosures not required.

health end points, such as cardiovascular and pulmonary, cancer, mental health, and integrating stress metrics and disrupted community metrics, could be incorporated into these studies to understand the link between environment, stress, and health.

DISCUSSION

After hearing the panel discussion, the individuals were asked to comment on suggested next steps. Dr. Goldstein noted that environmental health scientists need to be more involved in policy making and that they must be more emphatic about the importance of human health. This includes gathering data and documenting the fact that the public has a concern about their health and beginning a discussion with all stakeholders. Dr. Carey emphasized that there is a need to harness clinical data and combine this information with location. This approach could identify the hotspots for further investigations. He put forth that there is a need to move beyond anecdotal case reports to collecting systematic data that can be analyzed rigorously. Dr. Goldstein added the critical need for a National Institutes of Health study section to fund this type of research as opposed to enforcing new laws that would not inform researchers about the toxicological effects. Dr. Jackson added that funding needs to be front loaded so that when the inevitable clusters are identified (such as a cancer cluster, birth defect cluster, or neurological cluster) public health researchers will have initial data to launch investigations. He expressed concern that waiting until the clusters happen and trying to do a retrospective study is far too difficult and often inconclusive.

The discussion turned to the value of electronic health records and the problem that occupational history is often not recorded by health care professionals. Dr. Carey noted that even in the Geisinger Health System, employment data are collected, but not systematically, and therefore, there are gaps. He suggested that one way to fill the gaps is to use supplementary data collection modes. One participant noted that another weakness of the health records is that they are not the best source to assess the linkages between stress and health outcomes because some of these items are preclinical.

Dr. Donnelly discussed the role of industry and the need to create partnerships between vested parties. He sees a need to have a continued dialogue with the community and other stakeholders to define the research questions. He noted the need to have credible third parties to conduct research. Dr. Jackson and others noted that the Health Effects Institute and the Public Health Institute are good models that should be explored to begin research in this area. Dr. Donnelly added that Royal Dutch Shell is committed to safe operating practices, and they are also open to being informed by the research and dialogue in order to make changes. He indicated that there is also a need to continue to identify alternative chemicals that may offer better solutions and safer community-based solutions. Dr. Donnelly concluded that Royal Dutch Shell has called for transparency of the chemical composition of hydraulic fracturing fluid. The company participates in the fracfocus.org website where individuals can review the chemicals that are being used in their wells.

REFERENCE

Goldstein, B. D., J. Kriesky, and B. Pavliakova. 2012. Missing from the table: Role of the environmental public health community in governmental advisory commissions related to Marcellus Shale drilling. *Environmental Health Perspectives* 120(4):483–486.

9

Research Opportunities: Federal Representatives

This chapter presents a summary of a high-level discussion of environmental health and hydraulic fracturing from the perspective of the U.S. Environmental Protection Agency (EPA) and the research needs and opportunities to ensure that the public's health is protected. The chapter also includes a summary of perspectives of four federal agency representatives—from the National Institute of Environmental Health Sciences (NIEHS), the U.S. Geological Survey (USGS), the Occupational Safety and Health Administration (OSHA), and the Centers for Disease Control and Prevention (CDC)—on agency research activities. The presentations are followed by a summary of the discussion that occurred between panelists and members of the audience. The dialogue covered a broad range of topics raised over the 2-day workshop.

ENVIRONMENTAL HEALTH AND HYDRAULIC FRACTURING

Bob Perciasepe, M.P.A.
Deputy Administrator, U.S. Environmental Protection Agency

Bob Perciasepe stated that while the EPA plays a broad role in environmental protection and natural resources management, its scope of work goes well beyond wildlife and endangered species to the EPA's priority in environmental public health. The agency is guided by science and laws to ensure that the agency implements protections to combat air pollution, water pollution, unsafe drinking water, and soil contaminants. The EPA's work addresses many of the environmental challenges facing society today. In this approach of protecting human health, there will be cobenefits for wildlife and the broader natural resources, according to Mr. Perciasepe.

Many of the challenges facing the United States today are related to developing a sustainable and robust economy. Mr. Perciasepe believes that addressing these challenges are part of the EPA's mission, particularly in terms of the debate about whether a strong economy and environmental protection can work together. He stressed that the current U.S. economic conditions or any economic crisis is unrelated to having high environmental standards. The importance of having a clean, sustainable economy has been reinforced by President Obama in his 2012 State of the Union Address when he spoke about an economy built to last—one built on clean, cheaper energy.

Mr. Perciasepe emphasized the need for developing new energy sources and how the EPA reviews existing energy resources and also plays a proactive role in identifying new sources such as wind and solar. In other words, the EPA is prioritizing safe and effective ways to maximize the nation's oil and gas resources.

Similar to previous speakers, Mr. Perciasepe reiterated that natural gas will play a positive role in the nation's near-term energy policy in order to have a steady supply and ensure energy security. However, the EPA has heard the many concerns raised about human health implications of oil and gas development. It is clear that those concerns need to be addressed with care, he said.

Natural gas, from a pollution standpoint, releases less carbon dioxide than other forms of energy. While noting this, Mr. Perciasepe cautioned that these resources have to be developed safely and responsibly. Natural gas development has the ability to create jobs and provide fuel for cleaner transportation. In moving toward safe ways to effectively explore and develop U.S. natural gas resources, the EPA has allocated 45 million dollars toward ensuring a coordinated, interagency, research effort. Mr. Perciasepe also discussed the new executive order that is requiring agencies to be more coordinated in looking at all of the issues related to natural gas development, including hydraulic fracturing. The EPA, Department of Energy, and Department of the Interior have signed a memorandum of understanding to coordinate the expenditure of funds on research into all the different aspects of natural gas development. Although Mr. Perciasepe feels that there is a need to take full advantage of these technologies, he also stressed the need to give Americans confidence and ensure that neither their health nor natural resources will be sacrificed in that process.

The EPA has completed standards to reduce harmful air emissions and pollution associated with natural gas production. Under the Clean Air Act, new source performance standards are aimed at those emissions at the well head and also in some of the transport and pipeline storage systems. Mr. Perciasepe stated that these standards are important because as the wells are being prepared for natural gas production, they can emit volatile organic compounds (VOCs). VOCs contribute to ozone production, which results in the development of ozone problems. The wells can also emit chemicals, such as benzene and hexane, depending on the composition of the natural gas, which have potential carcinogenic effects. The EPA is focused not only on reducing VOCs from that process, but also reducing methane emissions, a greenhouse gas.

Regulations are based on input from the public, industry, environmental organizations, public health organizations, and state regulators. Mr. Perciasepe stated that these newer regulations are good examples of how natural gas resources can be developed while protecting public health and natural resources. The additional costs resulting from these regulations will be offset by the increase in natural gas captured; according to Mr. Perciasepe, an estimated $19 million will be saved annually by implementation of these regulations.

Mr. Perciasepe emphasized that safe water is a priority area of concern, including drinking water sources, the amount of water being used in the hydraulic fracturing process, and underground chemical injection control. The EPA has begun a study of the impact of hydraulic fracturing on drinking water resources, both groundwater and surface water. Mr. Perciasepe feels that when it is completed, the study will help move forward some of the scientific uncertainties related to water and hydraulic fracturing. Further, it will equip both the EPA and policy makers with evidence to inform decision making.

Mr. Perciasepe noted that the water used in hydraulic fracturing is recycled and reused, but eventually, the water will be disposed in a sewage treatment plant. Current EPA regulations do not allow for hydraulic fracturing fluids to be discharged into surface waters. The EPA is in the process of developing guidelines to regulate what quality level must be achieved before the water can be discharged into other treatment systems. Minimal disruption is the intent, according to Mr. Perciasepe.

The EPA is completing guidelines for underground injection control permitting groups in regional offices as well as some of the state agencies that have primacy for the underground injection control program on what practices should be used if any diesel fuel is used in the fracturing fluids. To the extent that these components of a fracturing fluid are not exempted under federal law, Mr. Perciasepe stressed that the EPA wants to ensure that there are proper guidelines in place if and when they use.

One common misconception he mentioned is that the EPA is standing in the way of natural gas and oil and natural resource development. Mr. Perciasepe noted that since 2008, natural gas and oil production in the United States has actually increased. Further, crude oil production in 2010 was higher than it was in any year in the previous decade. Production is occurring, but the EPA is committed to minimizing environmental and public health impacts.

Mr. Perciasepe reiterated that environmental and public health protection has a solid history to guide the approaches for the future. Forty years ago, air pollution in most cities in the United States was visible; water was visibly polluted, and other environmental hazards were clear. He noted that considerable progress has been made, but the public needs to understand that the challenges today are not quite as obvious as some of the challenges from 40 years ago. Today's challenges include climate change, existing pollutants, and the nation's aging infrastructure. Mr. Perciasepe is optimistic that the country can address these challenges and move forward and make great progress. He noted that most of the evidence shows that the country can protect the environment along with creating robust economic growth. He used the example that fuel economy standards for automobiles reduce greenhouse gas emission (6 billion metric tons of carbon pollution), other VOCs, and fine-particle pollution. At the same time, the standards improve national security efforts and energy security plans. He noted that the United States could save more than a trillion dollars in gasoline costs as a result of lower gasoline demand from the automobile and light-duty truck sector.

He stressed that although it is important to reduce demand on oil and develop alternative fuels, there is no single solution to all of these different mixes of energy issues in the United States. An interconnected, resilient energy system that supports policies that build a sustainable energy economy in the United States will protect the environment and public health.

PERSPECTIVES ON AGENCY RESEARCH ACTIVITIES

John M. Balbus, M.D., M.P.H.
Senior Advisor for Public Health,
National Institute of Environmental Health Sciences
National Institutes of Health

John M. Balbus described the NIEHS as a complex research institute with many programs that can help play a role in determining health implications of hydraulic fracturing. The institute is primarily a biomedical research institute with the majority of the funding activities being dedicated to investigator-initiated proposals. As an institute of the National Institutes of Health (NIH), Dr. Balbus noted that there is limited capacity to initiate new activities. The NIEHS is looking at existing mechanisms of support and ensuring that they provide the right mechanisms to start addressing and answering the kinds of questions raised during this workshop.

The NIEHS's intramural laboratory programs include the National Toxicology Program,[1] a freestanding clinical research unit, and extramural funding. Within the extramural program, Time Sensitive Grants, is a special small-grant-making program, Mechanism for Time-

[1] See http://ntp-server.niehs.nih.gov (accessed May 30, 2013).

Sensitive Research Opportunities in Environmental Health Sciences,[2] explicitly set up to address emerging issues and collect data in a relatively short time frame. A second existing mechanism is the Research to Action program,[3] which translates basic science into public health action. Dr. Balbus described recent funding announcements regarding cumulative stressors and community exposures in addition to environmental exposures.

The NIEHS has a number of environmental health science centers throughout the country that focus on topics such as breast cancer and children's environmental health. Each of the centers is required to have a community-oriented core consisting of community outreach, educational activities, and partnerships with community groups, and local and state decision makers. Some of the environmental health centers are located in areas that are being affected by hydraulic fracturing, according to Dr. Balbus. The directors are starting discussions and webinars and developing proposals to address some of the issues raised regarding the safety of the process.

Another avenue for involvement according to Dr. Balbus is the Workers' Training Program,[4] which was created to help train hazardous waste workers and emergency response workers on environmental health issues. He described it as a very robust network with partnerships between academic groups, educators, and labor unions. The network is a resource during emergency response periods.

The last program that Dr. Balbus detailed is the National Toxicology Program, a joint program between the Food and Drug Administration, the National Institute for Occupational Safety and Health (NIOSH), and the agencies of NIH. It exists to serve both the agencies and the public in evaluating the toxicity of substances, mixtures, and exposures of a whole variety of kinds from naturally occurring to synthetic materials. One of the components the National Toxicology Program is active investigation of the Tox21 high-throughput screening program—a program not set up to do in-depth toxicological evaluation, according to Dr. Balbus, but to screen large numbers of substances at once. The National Toxicology Program accepts public and self-generated nominations of substances for study. The group is reviewing hydraulic fracturing chemicals and seeking to determine the most appropriate mixtures. One large challenge that Dr. Balbus highlighted is the lack of information about what the public is being exposed to and what are the mixtures of greatest concern.

---

[2] See http://grants.nih.gov/grants/guide/pa-files/PAR-13-136.html (accessed May 30, 2013).

[3] See http://www.niehs.nih.gov/research/supported/dert/sphb/programs/peph/prog/rta (accessed May 30, 2013).

[4] See http://www.niehs.nih.gov/research/supported/dert/wet/index.cfm (accessed May 30, 2013).

In conclusion, Dr. Balbus noted a number of epidemiologic needs, including

- baseline health status of affected communities,
- site characterization and identification of highest-priority substances and mixtures to undergo toxicologic testing, and
- coordination between levels of government and agencies to respond to communities, obtain samples and health data, and communicate results.

Suzette M. Kimball, Ph.D.
Associate Director for Geology, U.S. Geological Survey
U.S. Department of the Interior

Suzette M. Kimball provided background to the audience on the USGS. The agency does not have regulatory or resource management responsibilities. Its mission is to produce objective, unbiased science and thus is uniquely positioned to support the President's executive order that emphasizes the need for coordinated science and coordinated activities to support the decisions surrounding environmental health issues potentially related to shale gas extraction. The USGS's key role includes supporting safe and responsible development of both conventional and unconventional domestic natural gas resources and in understanding the global distribution, global extent, and global access to energy resources. Dr. Kimball noted that the agency's work has for many years looked at energy resource assessments, hazards associated with seismic activity, energy development-associated risks, hydrogeologic investigations, and most recently, environmental health. She proposed that the need is paramount for the first three in order to understand the geologic framework and the hydrologic and geologic conditions that are associated with the extraction of these resources in order to be able to make informed decisions about environmental health.

Dr. Kimball expanded by saying that for all research areas in hydraulic fracturing, interagency collaborations will be important. This includes federal, state, and local agencies collaborating with non-governmental organizations and industry. The memo of understanding recently signed by the USGS, the Department of Energy, and the EPA will be able to better align federal resources so that the agencies bring all of the resources that the federal government has to bear on understanding the issues of concern.

Dr. Kimball described a number of science priorities for the USGS, including

- resource assessments for both conventional and unconventional resources and the related environmental impacts (both ecological and human components);

- water use and impact on supplies for humans and ecosystems;
- Impacts of produced and flowback waters on aquatic life in receiving water bodies;
- geochemistry and toxicity of fluids from shale gas wells;
- habitat destruction and forest fragmentation;
- assessments of the geographic footprint of extraction and related activities; and
- induced seismicity.

The agency, according to Dr. Kimball, is focusing its interdisciplinary science to address the growing complex questions around potential environmental health impacts in several key areas, such as resource extraction, production storage and transmission activities, and life-cycle assessments of related issues including waste management activities. To address these large-scale activities, several projects have been started. Dr. Kimball described the projects as a series of resource assessments both nationally and globally on tight gas, shale gas, tight oil, and coal bed methane. In addition to continuing its nationwide stream gauging and water availability studies, the USGS is completing water quality sampling and monitoring, targeting studies, and tailoring the existing network so that environmental health issues can be addressed. As mentioned earlier, the USGS is planning to sample and analyze flowback water for natural radiation. The agency scientists are also developing laboratory methods to measure chemicals in a wide range of fluids, including the fluids produced in hydraulic fracturing. Other research activities include using groundwater flow modeling to predict the fate of injected fluids, documenting landscape changes using specific satellite imagery to understand implications for wildlife, and induced seismicity as a result of wastewater and fluid injection subsequent to a fracturing process.

Dr. Kimball concluded by stating that all of those are efforts that will provide some of the baseline studies to inform environmental health decision making.

David M. Michaels, Ph.D., M.P.H.
Assistant Secretary of Labor,
Occupational Safety and Health Administration
U.S. Department of Labor

David M. Michaels commenced his presentation by noting that workers are an integral part of the safe use and the safe production of unconventional resources and at the same time, they are vulnerable to potential adverse events associated with hydraulic fracturing.

Oil and gas drilling in general is one of the most hazardous occupations in the United States. Over the years the fatality rate has been about seven times higher than the rate for all U.S. workers (CDC, 2012) and that is before the proliferation of hydraulic fracturing.

Shale gas extraction has many of the same hazards associated with this industry, but also introduces new hazards previously unseen to this extent in the petroleum industry. At times when new technologies are put into place, new hazards are seen; although the hazards associated with shale gas extraction are not new—silica exposure, diesel exposure, falls, and motor vehicle incidents—it is the constellation of these hazards that is unique. Motor vehicles, for example, are the major cause of fatalities involved in upstream oil and gas production. The task of bringing in millions of gallons of water and hundreds of thousands of tons of sand to shale gas extraction sites requires large numbers of vehicles to travel off-road and that has resulted in increased fatalities.

There are multiple tools to address workplace hazards. OSHA works closely with employers and industry to reduce workplace hazards. There are a number of large employers and large companies that are involved in shale gas extraction who are very committed to safety. Many of these companies extend safety standards down to all their contractors. A number of oil and gas companies formed the Service, Transmission, Exploration, and Production Safety Network (STEPS).[5] This network promotes safety, health, and environmental improvement in the exploration and production of oil and gas. Network meetings emphasize educating companies and contractors about the safety issues. However, new people enter this industry all the time who may not necessarily have the skills or the commitment to worker safety, and so, education should be a continuous effort.

Another approach that OSHA uses is to work jointly with other agencies that are involved earlier in the shale gas extraction process, for example, working with agencies responsible for permitting. It is critical to be involved early before drilling occurs and during the drilling process. Once the drilling is complete and the product is being extracted, the risks, injuries, and fatalities are very low. Dr. Michaels further noted that there is a mosaic of federal and state agencies that have different regulatory responsibilities with some overlap, but some gaps. The usual model of oversight typically results in one visit in the course of a long period of production and that often does not allow enough time to see the hazards. Working jointly with other agencies helps expand the opportunities to identify and address potential hazards.

---

[5] The Network changed its name from the South Texas Exploration and Production Safety Network (OSHA, 2007).

Christopher J. Portier, Ph.D.
Director, National Center for Environmental Health and Agency for Toxic Substances and Disease Registry
Centers for Disease Control and Prevention

Christopher J. Portier stated that CDC is America's public health agency for research and response to ongoing and emerging public health issues. The Agency for Toxic Substances and Disease Registry (ATSDR) and the National Center for Environmental Health (NCEH) have a number of ways to contribute to the understanding of the health implications of this new technology. One way is to study the potential toxicity of the compounds used in shale gas extraction. ATSDR maintains toxic profiles that are regarded internationally as some of the best reviews of the toxicological literature, and staff at ATSDR and NCEH are working with the NIEHS and with the EPA to evaluate these compounds. A second way is to directly evaluate human effects from shale gas extraction. ATSDR and NCEH have human health study groups dedicated to understanding health impacts from environmental exposures; they have been asked to identify what information is needed to clarify concerns and reduce any public health effects from shale gas extraction. A third way to contribute to the understanding of health effects of shale gas extraction was initiated by Congress. Congress specifically asked the Environmental Health Tracking Network at NCEH to develop baseline community health data that will eventually allow communities to monitor the impact of current and future drilling sites on the health of individuals living nearby. NCEH currently has a program on the safety of unregulated water. This program may need to be extended and linked with environmental public health tracking to improve surveillance in areas where shale gas extraction is prevalent. Finally, the medical education group at NCEH can develop guidance for health care professionals to enhance their understanding of the effects of some of the hazards seen with shale oil extraction, such as noise and stress, on overall health status.

In the quest to protect people and save lives, Dr. Portier said, it is clear that we should be working closely with our colleagues in industry to advance public health through identifying ways to help improve shale gas extraction practices. The panel was asked what makes shale gas extraction a unique concern. There are several characteristics of shale gas extraction that have raised questions among health professionals. The chemical constituents are not well characterized, and toxicity data on many of them are limited. A number of questions exist about the mobilization of naturally occurring substances such as methane and heavy metals. The good news, he said, is that many of these issues do have solutions and can be addressed through scientific study. There are opportunities where past experience can be used to reduce concerns almost immediately. The physical hazards associated with drilling and the effects of noise and air pollution at drilling operations are not new to shale gas extraction, and the industry has developed ways to limit impacts from these hazards. Best practices, if applied across the board, will have an immediate impact, for example, in the areas of waste handling, treatment, and long-term storage. Dr. Portier concluded that the way forward is for all who are interested in environmental health—local, state, tribal, and federal government agencies, industry, national news media, not-for-profit organizations, and certainly the affected communities—to work together.

DISCUSSION

During the discussion with the speakers, the panel members responded to a number of questions about the public health response, the monitoring of environmental exposures, and the local public health needs.

It was noted by one member of the audience that the United States was not the only country with shale gas and asked if the United States is aware of other countries' activities in this area. Dr. Kimball noted that the European Union is interested in the topic as are France, Great Britain, Russia, and the Ukraine. She noted that there has not been an assessment of the global reserves of shale gas but the geographic survey directors globally are meeting to begin to understand those reserves.

During the discussion, it was noted that different agencies have responsibility for different aspects of the hydraulic fracturing process and that front-line public health officials may not know which agency to call. Dr. Portier noted that NCEH's health education unit is putting together materials that will be shared with the states. However, he noted that NCEH is still in the information-gathering phase and will need to collect additional information before it can provide leadership and materials to guide this effort. Dr. Balbus noted that the NIEHS's environmental health science centers have a network—Partnerships for Environmental Public Health—of scientists, community members, educators, health care providers, public health officials, and policy makers who share the goal of increasing the impact of environmental public health research at the local, regional, and national levels. This network communicates between the NIEHS and stakeholders in the communities, including community groups and state and local public health officials. As the centers in the affected areas begin to develop education and training activities, this information will be a resource that will be available nationwide on the Internet.

Another audience member commented that the health response has been fragmented across the federal, state, and local governments. Dr. Michaels noted that the original interagency working group did not include OSHA, but this has changed. He said that there was a commitment on behalf of all the agencies to work more cooperatively and creatively to address these very tough issues. Dr. Goldman noted that it comes back to a recurring theme of public health needing to be more involved. At the same time, she noted that health agencies are often reluctant to accept responsibility for something that they feel should be the responsibility of the EPA or OSHA. Dr. Goldman further stated that the scientific community that is concerned about health and medicine needs to ensure that representatives in state health departments and the Department of Health and Human Services understand that environmental health is about health. Health goes beyond health care payments, health care reform—health includes the environment, but it is a dimension of health that is often overlooked.

Another audience member noted that an unprecedented number of workforce reductions have occurred at the state and local levels which makes responding to challenges such as hydraulic fracturing difficult, if not impossible. The audience member asked how can the public health infrastructure of the country be maintained or rebuilt. Dr. Portier noted that the CDC is committed to trying to maintain and hold together the strength of the public health force of the United States. In the past few years, state and local governments have lost approximately 700,000 jobs, and approximately 200,000 of those are in health and environmental areas. Dr. Balbus added that environmental health has not been as effective as other entities in making the economic case for rebuilding the public health infrastructure. He suggested that there is a research and science need to build environmental health economics in order to make the economic case for an investment in the environmental health infrastructure.

Another audience member noted that currently many of the agencies and organizations that are involved in addressing environmental health concerns are in a reactive mode. The audience member questioned whether there were opportunities for federal agencies to develop voluntary guidelines for states on issues such as monitoring wells prior to their installation. Dr. Portier noted that the CDC would be willing to work with other agencies, such as the EPA, to develop exemplary or model guidelines for states to consider.

An audience member raised the concern about the potential health effects of radon that are associated with the natural gas that is extracted and that will eventually go to the consumer. As the gas travels to the consumer, it will be plating out in the distribution lines with polonium-210 and lead-210, both of which are solids that make those pipes radioactive, and maintenance workers will be exposed during the process. Further, he noted, radioactivity levels in the produced gas in these various fields has not been examined in some time. There were EPA studies in the 1970s of radon in gases from shale deposits across the country and USGS data from the Marcellus Shale in the 1980s, but more recent studies are lacking. The audience member asked the panel if agencies were concerned about radon in the end product. Dr. Kimball responded that the USGS is undertaking a set of studies of naturally occurring radionuclide elements. It is known that within the Piedmont areas of the East Coast, radon is an issue, but it is an issue for a number of reasons, not just shale gas extraction. The state geological surveys are in the process of completing state geologic maps. As they begin to look beyond surface geology, there is an opportunity to look at this issue. Those studies are under way.

Finally, the panel was asked to comment on the testing of acute, short-term exposures versus low-level chronic exposures, for example, the low-level chronic exposures of farmers who leased out their land for hydraulic fracturing or homeowners who are living 100 feet from a compressor station and live with these emissions daily. The audience member noted that there has been remarkably little air and water testing in the U.S. gas fields to date, and the available testing efforts have shown exposures at "safe" levels, which is disheartening for people experiencing a multiplicity of health symptoms at these levels. The audience member questioned whether these standards needed to be changed. Dr. Portier responded by explaining that NCEH's toxicology profiles are guidance values for people to use in deciding whether to act in a particular situation; but they are not standards. There are three different guidance values for each level of exposure: acute (exposures of less than a few hours to 1 week), medium (up to 1 year), and chronic exposures (greater than 1 year). For the guidance, he noted that what is allowed for acute exposure is greater than what is allowed for medium-term or chronic exposure. Dr. Portier noted that many of the decisions that are made are based upon the acute exposure value.

REFERENCES

CDC (Centers for Disease Control and Prevention). 2012. Oil and gas extraction. Occupational safety and health risks: Fatalities. http://www.cdc.gov/niosh/programs/oilgas/risks.html (accessed May 30, 2013).

OSHA (Occupational Safety and Health Administration). 2007. Annual Alliance Report: South Texas Exploration and Production Safety Network (STEPS). http://www.osha.gov/dcsp/alliances/regional/reg6/steps_annualreport2007-2008.html (accessed May 30, 2013).

A

Agenda

WORKSHOP ON THE HEALTH IMPACT ASSESSMENT OF NEW ENERGY SOURCES: SHALE GAS EXTRACTION

Sponsored by
Roundtable on Environmental Health Sciences, Research, and Medicine

April 30–May 1, 2012
House of Sweden
2900 K Street NW, Washington, DC

Workshop Goals and Objectives

1. Use shale gas extraction to explore the health impacts of emerging energy technologies.
2. Application of health impact assessments to identify ways to mitigate adverse health effects; state of the science.
3. Identify direct and indirect health risks and solutions from a cradle-to-grave approach. Draw from analogous conditions when data are incomplete.
4. Identify vulnerable populations and stakeholders.
5. Describe research questions, data sources, data gaps, and how to address uncertainty. Identify opportunities to draw from insights from similar and well-characterized operations conducted in different regions of the country or world.
6. Discuss next steps for stakeholders.

Day 1—April 30

8:00 a.m.	**Welcome** Harvey V. Fineberg, M.D., Ph.D. President, Institute of Medicine
8:10 a.m.	**Opening** Lynn R. Goldman, M.D., M.P.H. Vice Chair, Roundtable on Environmental Health Sciences, Research, and Medicine Dean, School of Public Health, George Washington University
8:20 a.m.	**Charge of the Workshop** Christopher J. Portier, Ph.D. Director, National Center for Environmental Health and Agency for Toxic Substances and Disease Registry Centers for Disease Control and Prevention
8:35 a.m.	**Hydraulic Fracturing: Accessing Shale and Tight Gas** David Cole, M.S. Regional Discipline Leader—Production Technology/Chemistry Shell Upstream Americas
8:55 a.m.	**Health Impact Assessment for Shale Gas Extraction** Aaron Wernham, M.D., M.S. Project Director, Health Impact Project Pew Charitable Trusts
9:15 a.m.	**Discussion** *Moderator*: Lynn R. Goldman, M.D., M.P.H.
9:35 a.m.	**BREAK**

9:50 a.m.	**Setting the Stage for Health: Why Fracking, Why Now?: Identifying Key Issues That Are Unique and Potential Priorities** Linda A. McCauley, Ph.D., R.N., FAAN, FAAOHN Dean, Nell Hodgson Woodruff School of Nursing Emory University
10:10 a.m.	**The Geographic Footprint** Michael Focazio, Ph.D. Assistant Program Coordinator Toxic Substances Hydrology Program U.S. Geological Survey
10:30 a.m.	**Assessing the Perceived and Real Environmental Consequences of Shale Gas Development** Charles G. Groat, Ph.D. John A. and Katherine G. Jackson Chair in Energy and Mineral Resources Department of Geological Sciences Professor of Geological Sciences and Public Affairs Lyndon B. Johnson School of Public Affairs The University of Texas at Austin
10:50 a.m.	**Discussion** *Moderator*: Anne M. Sweeney, Ph.D. Professor, Department of Epidemiology and Biostatistics School of Rural Public Health Texas A&M University
11:20 a.m.	**NIOSH Field Effort to Assess Chemical Exposures in Oil and Gas Workers** Eric J. Esswein, M.S.P.H. Senior Industrial Hygienist National Institute for Occupational Safety and Health Centers for Disease Control and Prevention
11:40 a.m.	**Discussion** *Moderator*: Linda A. McCauley, Ph.D., R.N., FAAN, FAAOHN

12:05 p.m. **LUNCH BREAK**

1:00 p.m. **Community Impacts of Natural Gas Development and Human Health**
Roxana Witter, M.D., M.S.P.H., M.S.
Assistant Research Professor
Environmental and Occupational Health
Colorado School of Public Health

1:20 p.m. **Economic and Community Impacts of Gas Shale in Pennsylvania**
Timothy Kelsey, Ph.D.
Professor of Agricultural Economics
State Program Leader, Economic & Community Development
The Pennsylvania State University

1:40 p.m. **Discussion**
Moderator: Linda A. McCauley, Ph.D., R.N., FAAN

2:00 p.m. **Potential Air Quality Impacts of the Development and Production of Marcellus Shale Gas**
Allen Robinson, Ph.D.
Professor, Mechanical Engineering, and Engineering and Public Policy
Carnegie Mellon University

2:20 p.m. **Air Quality Impacts of Natural Gas Operations in Texas**
Michael Honeycutt, Ph.D.
Toxicology Division Director
Texas Commission on Environmental Quality

2:40 p.m. **Air Pollution Exposure and Risk Near Unconventional Natural Gas Drill Sites: An Example in Garfield County, Colorado**
John Adgate, Ph.D., M.S.P.H.
Professor and Chair, Department of Environmental & Occupational Health
Colorado School of Public Health

3:00 p.m. **Air Respondent**
Bernard D. Goldstein, M.D.
Professor Emeritus, Department of Environmental and Occupational Health, Graduate School of Public Health
University of Pittsburgh

3:05 p.m. **Discussion**
Moderator: Richard A. Fenske, Ph.D., M.P.H.
Associate Chair and Professor
Department of Environmental and Occupational Health Sciences, School of Public Health
University of Washington

3:35 p.m. **BREAK**

3:50 p.m. **Potential Impacts of Hydraulic Fracturing on Water Resources**
Deborah L. Swackhamer, Ph.D., M.S.
Professor, Co-Director of the Water Resources Center
Division of Environmental Health Sciences
University of Minnesota School of Public Health

4:10 p.m. **Hydraulic Fracturing, Water Resources, and Human Health**
Robert B. Jackson, Ph.D., M.S.
Nicholas Chair of Global Environmental Change
Nicholas School of the Environment
Professor, Department of Biology
Duke University

4:30 p.m. **EPA Study Plan on the Potential Impacts of Hydraulic Fracturing on Drinking Water Resources: Approach to Study Potential Health Impacts**
Jennifer Orme-Zavaleta, Ph.D.
Director, National Exposure Research Laboratory
Office of Research and Development
U.S. Environmental Protection Agency

4:50 p.m.	**Discussion** *Moderator*: James K. Bartram, Ph.D. Professor of Environmental Sciences and Engineering Director of the Water Institute Gillings School of Global Public Health University of North Carolina at Chapel Hill
5:20 p.m.	**ADJOURN FOR THE DAY**

Day 2—May 1, 2012

8:00 a.m.	**Welcome Back** Lynn R. Goldman, M.D., M.P.H.
8:15 a.m.	**Sustainable Energy for All?** Steven Hamburg, Ph.D., M.F.S. Chief Scientist, Environmental Defense Fund
8:35 a.m.	**Sustainable Energy for All: Ensuring Health Throughout the Energy Production and Use Life Cycle** Daniel S. Greenbaum, M.S. President, Health Effects Institute
8:55 a.m.	**Discussion**
9:20 a.m.	**Panel I: Research Community** Each panelist will give a 5-minute presentation followed by discussion. **Assessment of the Science and Next Steps** • What evidence links adverse health effects and hydraulic fracturing? • What can be done to minimize adverse health effects as the technology evolves (e.g., best practices)? • Where is there uncertainty (appropriate metrics)? • What are the next steps for stakeholders?

	Moderator: Richard J. Jackson, M.D., M.P.H. Professor and Chair, Department of Environmental Health Sciences, School of Public Health University of California, Los Angeles
	David Carey, Ph.D. Director, Weis Center for Research Geisinger Health Center
	Rob Donnelly, M.B.Ch.B., MFOM Vice President of Health Royal Dutch Shell
	Bernard D. Goldstein, M.D. Professor Emeritus, Department of Environmental and Occupational Health, Graduate School of Public Health University of Pittsburgh
	Roxana Witter, M.D., M.S.P.H. Assistant Research Professor Environmental and Occupational Health Colorado School of Public Health
10:00 a.m.	**Discussion**
10:30 a.m.	**BREAK**
10:50 a.m.	**Environmental Health and Hydraulic Fracturing** Bob Perciasepe, M.Pl., M.P.A. Deputy Administrator U.S. Environmental Protection Agency
11:05 a.m.	**Discussion**
11:15 a.m.	**Panel II: Federal Representatives** Each panelist will give a 5-minute presentation followed by discussion.

Identification of Key Research and Regulatory Needs

- From your agency's perspective, what do you see as clear opportunities for action?
- What is your agency's critical advantage for addressing questions about shale gas extraction?
- What makes shale gas extraction unique? What are opportunities to draw from insights from similar and well-characterized operations conducted in different regions of the country or world?
- How can local, state, and federal institutions coordinate to make efficient progress?
- Are there other partners that can help advance our knowledge?

Moderator: George M. Gray, Ph.D.
Professor, Department of Environmental and Occupational Health
Director, Center for Risk Science and Public Health
School of Public Health and Health Services
The George Washington University

John M. Balbus, M.D., M.P.H.
Senior Advisor for Public Health
National Institute of Environmental Health Sciences
National Institutes of Health

Suzette M. Kimball, Ph.D.
Associate Director for Geology, U.S. Geological Survey
U.S. Department of the Interior

David M. Michaels, Ph.D., M.P.H.
Assistant Secretary of Labor
Occupational Safety and Health Administration
U.S. Department of Labor

Christopher J. Portier, Ph.D.
Director, National Center for Environmental Health and Agency for Toxic Substances and Disease Registry
Centers for Disease Control and Prevention

12:00 p.m.	**Discussion**
12:30 p.m.	**Closing** Frank Loy, L.L.B. Chair, Roundtable on Environmental Health Sciences, Research, and Medicine U.S. Representative to the 66th Session of the General Assembly of the United Nations
12:45 p.m.	**ADJOURN**

B

Speaker Biosketches

John Adgate, Ph.D., M.S.P.H., is chair of the Department of Environmental & Occupational Health at the Colorado School of Public Health. His research focuses on improving exposure assessment in epidemiologic studies—studying the factors that affect the health and illness of entire populations—by documenting the magnitude and variability of human exposure to air pollutants, pesticides, metals, and allergens. Some of his research projects have included evaluating methods that might help to reduce lead poisoning in the home, outcomes of long-term exposure to indoor and outdoor air pollutants, and a controlled trial to test an allergen reduction intervention in inner-city residences. Dr. Adgate has a Ph.D. in environmental health from the University of Medicine and Dentistry of New Jersey–Robert Wood Johnson Medical School and Rutgers University, an M.S. in public health from the University of North Carolina at Chapel Hill, and a B.S. in biology from Calvin College.

James K. Bartram, Ph.D., is a professor of environmental sciences and engineering in the Gillings School of Global Public Health of the University of North Carolina (UNC) at Chapel Hill and founding director of the Water Institute at UNC. He has worked in diverse areas of public health and disease prevention, especially in relation to environment and health and water supply and sanitation. From 1998 to 2009, he worked at the World Health Organization's Headquarters, leading the Water, Sanitation, Hygiene, and Health Unit and the Unit for Assessing and Managing Environmental Risks to Health. Dr. Bartram was awarded the International Water Association Grand Award in 2004 for international leadership in development and application of evidence-based policy and good practice. He is an honorary professor at the University of Wales at Aberystwyth and a visiting professor at the Universities of Bristol and Surrey, United Kingdom. Dr. Bartram is author of more than 60 academic papers and more than 40 book chapters, and editor of about 25 books, including aspects of global monitoring, water supply, sanitation, and pollution. He received a Ph.D. in environmental and public health and a B.Sc. in microbiology from the University of Surrey.

Linda S. Birnbaum, Ph.D., DAPT, ATS, is director of the National Institute of Environmental Health Sciences (NIEHS) of the National Institutes of Health (NIH), and National Toxicology Program (NTP). As NIEHS and NTP director, Dr. Birnbaum oversees a budget that funds multidisciplinary biomedical research programs, prevention, and intervention efforts that encompass training, education, technology transfer, and community outreach. The NIEHS supports more than 1,000 research grants. A board-certified toxicologist, Dr. Birnbaum has served as a federal scientist for nearly 32 years. Prior to her appointment as the NIEHS and NTP director in 2009, she spent 19 years at the U.S. Environmental Protection Agency (EPA) where she directed the largest division focusing on environmental health research. Dr. Birnbaum started her federal career with 10 years at the NIEHS, first as a senior staff fellow at the National Toxicology Program, then as a principal investigator and research microbiologist, and finally as a group leader for the Institute's Chemical Disposition Group. Dr. Birnbaum has received numerous awards and recognitions. She was elected to the Collegium Ramazzini, received an honorary doctor of science from the University of Rochester, and a Distinguished Alumna Award from the University of Illinois. In October 2010, she was elected to the Institute of Medicine of the National Academies. Her awards include the Women in Toxicology Elsevier Mentoring Award, the Society of Toxicology Public Communications Award, EPA's Health Science Achievement Award and Diversity Leadership Award, and 12 Science and Technology Achievement Awards. She is the author of more than 700 peer-reviewed publications, book chapters, abstracts, and reports. Dr. Birnbaum received her M.S. and Ph.D. in microbiology from the University of Illinois at Urbana-Champaign.

David J. Carey, Ph.D., is associate chief research officer, director, and senior scientist of the Sigfried and Janet Weis Center for Research at the Geisinger Health System. Dr. Carey has extensive research experience in the areas of cellular and molecular biology, and is now extensively engaged in genomics research on vascular disease and other areas. He has served as a key player in the development of translational genomics research at Geisinger since 2004. Dr. Carey received his Ph.D. from St. Louis University.

David Cole, M.S., is regional discipline leader for production technology for Shell's Upstream Americas Business Unit. In this role, he oversees the activities of Shell's 200 production engineers in the Americas. Mr. Cole joined Shell in 1981 as a production engineer. He spent the next 10 years being responsible for conventional and unconventional fields in Shell's Onshore and Offshore businesses. Mr. Cole then moved to Shell's Bellaire Research Center where he was a member of a team positioning Shell for its entrance into the Deepwater. In 1993, Mr. Cole transferred back to a production engineering technical specialist before becoming a superintendent with engineering and operations responsibilities for drilling, completions, and well inter-ventions. Following the creation of the Americas Region in 2003, Mr. Cole became completion and wells services engineering manager for the region. Moving back into the technology arena, he then headed up the region's Technology Planning and Implementation team, with responsibility for developing new capabilities for the region. Prior to his current assignment, Mr. Cole was operations manager with responsibility for three of Shell's platforms in the Gulf of Mexico. He has a bachelor's degree in mechanical engineering from Mississippi State University and a master's degree in petroleum engineering from Louisiana State University.

Rob Donnelly, M.B.Ch.B., MFOM, grew up in Perth Scotland and studied Medicine at Edinburgh University. He joined the Royal Army Medical Corps and spent 6 years in a variety of roles, with the infantry in Northern Ireland and London, Army hospitals in Hong Kong and London, and finally a mobile armored field unit in the West of England. He completed training in family medicine before leaving the Army to train in occupational medicine at British Steel in South Wales. After 6 years in a variety of roles in the steel industry, he moved to Shell in Aberdeen, working in the offshore Exploration and Production sector. A move to Houston, Texas, followed, where for 4 years he led Shell Health Services in the Americas. This involved multiple business units in 34 countries. In March 2007 he moved to The Hague, Netherlands, to assume his current position as vice president health for Royal Dutch Shell. He is accountable for 650 staff in 130 countries where Shell has operational interests. A particular area of focus is operations and the potential impact on health of a community. His professional interests include fitness to work and environmental health. He has published a number of articles on occupational medicine and health and the workplace.

Eric J. Esswein, M.S.P.H., is a senior industrial hygienist with the National Institute for Occupational Safety and Health (NIOSH) Western States Office in Denver, Colorado. He conducts field-based research in oil and gas exploration and production. Esswein has been a commissioned officer in the U.S. Public Health Service since 1991 when he joined NIOSH as an industrial hygienist with the Hazard Evaluations and Technical Assistance Branch in Cincinnati, Ohio, before transferring to the NIOSH Denver Field Office in 1998. He earned a bachelor's degree in environmental health and toxicology from Huxley College of the Environment at Western Washington University and a master's degree in public health with an emphasis in industrial hygiene from the University of Utah.

Richard A. Fenske, Ph.D., M.P.H., is professor and associate chair of environmental and occupational health sciences at the University of

Washington (UW), and has served as director of the NIOSH-supported Pacific Northwest Agricultural Safety and Health Center since 1996. He is a core faculty member of the NIEHS-supported Center for Ecogenetics and Environmental Health. He also served as deputy director of the EPA/NIEHS-supported UW Center for Child Environmental Health Risks Research from 1996 to 2003, and director of the UW Field Research and Consultation Group from 1992 to 1996. Dr. Fenske currently serves on several federal advisory boards and committees: The EPA's primary advisory group, the Science Advisory Board; the National Academy of Sciences/Institute of Medicine (IOM) Committee to Review the Health Effects in Vietnam Veterans of Exposure to Herbicides; and the 16-member EPA Human Studies Review Board, which evaluates the science and ethics of studies involving human subjects. He is also on the editorial boards of the *Journal of Agricultural Safety and Health* and the *Journal of Exposure Science and Environmental Epidemiology.* Dr. Fenske teaches courses in the areas of environmental health risk assessment, environmental sampling and analysis, exposure science, and public health policy related to pesticide use. From 1984 to 1990, Dr. Fenske was assistant professor and then associate professor of environmental sciences at Rutgers University and the New Jersey Agricultural Experiment Station. He received his doctoral degree and master's in public health from the University of California (UC), Berkeley, in environmental health sciences. He was also awarded a master's degree in geography from UC Berkeley and a master's degree in comparative religion from Columbia University in New York. His bachelor's degree is in history from Stanford University.

Michael Focazio, Ph.D., received his doctorate from the University of Cincinnati in 1988, specializing in watershed modeling. After a short time as environmental scientist for the Interstate Commission on the Potomac River Basin, Dr. Focazio joined the U.S. Geological Survey (USGS) as a hydrologist in the Virginia Water Science Center. He served as the program coordinator for the USGS/National Park Service Water Quality Partnership for several years and is presently the associate coordinator for the Toxic Substances Hydrology Program in the Energy, Minerals, and Environmental Health Mission Areas. Dr. Focazio is an instructor in the Johns Hopkins University Advanced Academic Programs, Kreiger School of Arts and Sciences and is the past USGS liaison to the EPA's Office of Groundwater and Drinking Water. He is presently an appointed board member for the Water and Sanitation Authority of Fauquier County, Virginia.

Lynn R. Goldman, M.D., M.P.H., a pediatrician and an epidemiologist, is dean of and professor at the School of Public Health and Health Services at George Washington University (GWU). Prior to her move to GWU, Dr. Goldman was a professor of environmental health sciences at the Johns Hopkins University Bloomberg School of Public Health. Her areas of focus are children's environmental health research, public health preparedness, and environmental health policy. She had joint appointments in the Departments of Health Policy and Management and Epidemiology and in Emergency Medicine at the John Hopkins School of Medicine. From 1993 to 1998, Dr. Goldman served as Assistant Administrator for the EPA's Office of Prevention, Pesticides and Toxic Substances. In that position she was responsible for the nation's pesticide, toxic substances, and pollution prevention laws, with responsible for managing a number of complex regulatory and science issues. Her achievements included expanding the Toxics Release Inventory, reauthorizing the nation's pesticides laws (Food Quality Protection Act of 1996); and development of a framework for the regulation of biotechnology chemical and pesticide products. She led consensual processes that developed frameworks for testing of high-volume industrial chemicals and for identification of chemicals that disrupt endocrine systems. Between 1985 and 1993, Dr. Goldman served at the California Department of Health Services, most recently as head of the Division of Environmental and Occupational Disease Control. She led public health efforts to respond to emergencies such as earthquakes and unintentional releases of pesticides in communities. She conducted public health investigations on pesticides, childhood lead poisoning, and other environmental hazards. She has a B.S. from UC Berkeley, an M.P.H. from the Johns Hopkins University Bloomberg School of Public Health, an M.D. from UC San Francisco, and pediatric training at Children's Hospital, Oakland, California. She has served on numerous boards and expert committees, including the Committee on Environmental Health of the American Academy of Pediatrics and the CDC Lead Poisoning Prevention Advisory Committee. Dr. Goldman is a member of the IOM and vice chair of the IOM Roundtable on Environmental Health Sciences, Research, and Medicine.

Bernard D. Goldstein, M.D., is emeritus professor of environmental and occupational health and former dean of the University of Pittsburgh Graduate School of Public Health. He is a physician, board certified in internal medicine, hematology, and toxicology. Dr. Goldstein is author of more than 150 publications in the peer-reviewed literature, as well as numerous reviews related to environmental health. He is an elected member of the IOM and of the American Society for Clinical Investigation. His experience includes service as assistant administrator for research and development of the EPA, 1983–1985. In 2001, he came to the University of Pittsburgh from New Jersey, where he had been the founding director of the Environmental and Occupational Health Sciences Institute, a joint program of Rutgers University and Robert Wood Johnson Medical School. He has chaired more than a dozen National Research Council (NRC) and IOM committees primarily related to environmental health issues. He has been president of the Society for

Risk Analysis; and has chaired the NIH Toxicology Study Section, EPA's Clean Air Scientific Advisory Committee, the National Board of Public Health Examiners, and the Research Committee of the Health Effects Institute. He has also served as a member or chairperson of numerous national and international scientific advisory committees for government, industry, and environmental groups.

George M. Gray, Ph.D., is professor in the Department of Environmental and Occupational Health and director of the Center for Risk Science and Public Health at the GWU School of Public Health and Health Services (SPHHS). In both academic and policy-making settings, Dr. Gray has long been committed to the effective use of science to inform public health choices, and emphasizes the importance of communicating those choices effectively to citizens, journalists, and lawmakers. Prior to joining SPHHS in 2010, Dr. Gray served as assistant administrator for the EPA's Office of Research and Development and as the agency science advisor, promoting scientific excellence in EPA research, advocating for the continuing evolution of the agency's approach to analysis, and encouraging programs that provide academic research to support EPA's mission. His areas of focus included nanotechnology, ecosystem research, the influence of toxicology advances on testing and risk assessment, and sustainability. From 2005 to 2009, Dr. Gray was executive director of the Harvard Center for Risk Analysis, and a member of the faculty at the Harvard School of Public Health. In addition to teaching, he applied the tools of risk analysis to public health problems ranging from mad cow disease to pesticides in food to the risks and benefits of fish consumption. Dr. Gray received his doctor of philosophy and master of science in toxicology from the University of Rochester School of Medicine and his bachelor of science in biology from the University of Michigan.

Dan Greenbaum, M.C.P., joined the Health Effects Institute (HEI) as its president and chief executive officer in 1994. In that role, he leads HEI's efforts, supported jointly by EPA and industry, with additional funding from the Department of Energy, the Federal Highway Administration, the U.S. Agency for International Development, the Asian Development Bank, and foundations, to provide public and private decision makers in the United States, Asia, Europe, and Latin America with high-quality, impartial, relevant, and credible science about the health effects of air pollution to inform air quality decisions in the developed and developing world. Mr. Greenbaum has been a member of the NRC Board on Environmental Studies and Toxicology and vice chair of its Committee for Air Quality Management in the United States. He recently served on the NRC Committee on the Hidden Costs of Energy and serves currently on the NRC Committee on Science for EPA's Future. Mr. Greenbaum also chaired the EPA Blue Ribbon Panel on Oxygenates in Gasoline, which issued the report *Achieving Clean Air and Clean Water and EPA's*

Clean Diesel Independent Review Panel, which reviewed technology progress in implementing the 2007 Highway Diesel Rule. In May 2010, Mr. Greenbaum received the Thomas W. Zosel Outstanding Individual Achievement Award from the EPA for his contributions to advancing clean air. Mr. Greenbaum holds bachelor's and master's degrees from Massachusetts Institute of Technology in city planning.

Charles G. Groat, Ph.D., holds the John A. and Katherine G. Jackson Chair in Energy and Mineral Resources in the University of Texas (UT) Department of Geological Sciences and is director of the Center for International Energy and Environmental Policy and the Energy and Mineral Resources Graduate Program. He joined the Department of Geological Sciences in June 2005 after serving for 6-and-a-half years as director of the USGS, appointed by President Clinton and retained by President Bush. He also has faculty appointments in the Lyndon B. Johnson School of Public Affairs and the Department of Petroleum and Geosystems Engineering. At the USGS, he emphasized integrated scientific approaches to understanding complex natural systems and the use of this understanding in management decisions regarding these systems, an interest that continues at the university. His degrees in geology are from the University of Rochester (A.B.), University of Massachusetts (M.S.), and UT at Austin (Ph.D.).

Michael Honeycutt, Ph.D., is the director of the Toxicology Division of the Texas Commission on Environmental Quality (TCEQ). He has been employed by the TCEQ since 1996 and has managed the division of 14 toxicologists since 2003. His responsibilities include overseeing health effects reviews of air permit applications, overseeing the review of the results of ambient air monitoring projects, and overseeing the reviews of human health risk assessments for hazardous waste sites. Dr. Honeycutt spearheaded the updating of TCEQ's Effects Screening Levels (ESLs), or toxicity factors for chemicals. The current TCEQ ESL derivation procedure has been through two independent external scientific peer reviews and multiple rounds of public comment. Dr. Honeycutt serves as a technical resource for TCEQ management and staff on issues concerning air and water quality, drinking water contamination, and soil contamination. He also serves as an expert witness in public and state legislative hearings, participates in public meetings, and has conducted hundreds of media interviews. Dr. Honeycutt is an adjunct professor at Texas A&M University, has published numerous articles in the peer-reviewed literature, serves or has served on numerous external committees, and has provided invited testimony at congressional hearings.

Richard J. Jackson, M.D., M.P.H., is a professor and chair of environmental health sciences at UC Los Angeles. He has worked extensively on the impact of the environment on public health, and over the past decade much of his work has focused on how the built environment affects health. In 2004, he was co-author of *Urban Sprawl and Public Health*. Dr. Jackson is currently working on policy analyses of environmental impacts on health, from chemical body burdens to climate change to urban design. In addition, he is evaluating the effects of farming, education, housing, and transportation policies on health. Dr. Jackson chaired the American Academy of Pediatrics Committee on Environmental Health and recently served on the Board of Directors of the American Institute of Architects. He serves on the editorial boards of the *American Journal of Industrial Medicine*, *Environmental Research*, and *Public Health Reports*. He is a member of the Institute of Medicine Roundtable on Environmental Health Sciences, Research, and Medicine and of the NRC Committee on "Sustainable" Products and Services. Dr. Jackson earned his M.D. from UC San Francisco.

Robert Jackson, Ph.D., M.S., is the Nicholas Chair of Global Environmental Change and a professor in the Biology Department and Nicholas School of the Environment and Earth Sciences at Duke University. His research examines feedbacks between people and the biosphere, including studies of the global carbon and water cycles, biosphere–atmosphere interactions, and global change. He is currently director of Duke's Center on Global Change and Duke's Stable Isotope Mass Spectrometry Laboratory. In his quest for solutions to global warming, he also directs the new Department of Energy–funded National Institute for Climatic Change Research for the southeastern United States and codirects the Climate Change Policy Partnership, working with energy and utility corporations to find practical strategies to combat climate change. Dr. Jackson has received numerous awards, including the Murray F. Buell Award from the Ecological Society of America, a 1999 Presidential Early Career Award in Science and Engineering from the National Science Foundation, a fellow in the American Geophysical Union, and inclusion in the top 0.5 percent of most cited scientific researchers. Dr. Jackson's research has been covered in various newspapers and magazines, such as the *Boston Globe*, *New York Times*, *Washington Post*, *USA Today*, *Scientific American*, and *BusinessWeek*, and on national public radio, including the syndicated programs "Morning Edition," "All Things Considered," "Marketplace," "The Tavis Smiley Show," "The Next 200 Years," and "Earth and Sky" (for which he is a science advisor and scriptwriter). Dr. Jackson received his B.S. in chemical engineering from Rice University (1983). He worked 4 years for the Dow Chemical Company before obtaining M.S. degrees in ecology (1990) and statistics (1992) and a Ph.D. in ecology (1992) at Utah State University. He was a Department of Energy Distinguished Postdoctoral Fellow for Global Change at Stanford University and an assistant professor at the University of Texas before joining the Duke faculty in 1999.

Timothy Kelsey, Ph.D., is a professor of agricultural economics at The Pennsylvania State University. He conducts research on issues such as economic and community implications of Marcellus Shale, public finance and taxation, and land-use change. Through Penn State Cooperative Extension, Dr. Kelsey teaches workshops statewide to local government officials, citizens, and others interested in community issues. He has been at Penn State since 1991, and began actively working on Marcellus issues in 2008.

Suzette M. Kimball, Ph.D., is associate director for geology of the USGS. Dr. Kimball is the first woman to hold the position. Previously, she was director of the USGS's Eastern Region. Dr. Kimball provides executive leadership of USGS geologic investigations on the past, present, and future conditions of the Earth's environment, hazards, and resources. Specifically, she is responsible for basic earth science programs, including monitoring of worldwide earthquake hazards, geologic mapping of land and seafloor resources, the study of volcanic and landslide hazards, and research and assessments of mineral and energy resources. As director of the Eastern Region of the USGS, Dr. Kimball led multidisciplinary science programs in geology, hydrology, biology, and geography, covering the 26 U.S. states east of the Mississippi River, the District of Columbia, Puerto Rico, and the U.S. Virgin Islands. The USGS Eastern Region includes more than 2,600 employees in about 120 locations. She joined USGS in 1998 as Eastern Regional Executive for Biology. In that position, she built many partnerships, helped shape programs, and led the establishment of the USGS Florida Integrated Science Center. Dr. Kimball received her B.A. in english from the College of William & Mary, an M.S. in geology/ geophysics from Ball State University, and a Ph.D. in environmental sciences/coastal and oceanographic processes from the University of Virginia.

Linda A. McCauley, Ph.D., R.N., FAAN, FAAOHN, is the sixth dean of Emory University's Nell Hodgson Woodruff School of Nursing. She began her appointment in May 2009 after serving as the associate dean for research at the University of Pennsylvania's School of Nursing. Dr. McCauley holds a secondary appointment in the Rollins School of Public Health at Emory University and is internationally recognized for her scholarship in environmental and occupational health. She has devoted much of her distinguished career to identifying culturally appropriate interventions to decrease the impact of environmental and occupational health hazards for workers and young children. Dr. McCauley is currently leading two studies in Florida and Oregon with funding from NIH and CDC. Dr. McCauley is an elected member of the IOM of the National Academies. She also is a fellow of the American Academy of Nursing and the American Academy of Occupational Health Nurses. She has been widely published in the fields of nursing and environmental health. She is a sought-after speaker and has been featured in national publications and broadcasts including *Time*, *Business Week*, the *Atlanta Journal-Constitution*, National Public Radio, and the Weather Channel.

David M. Michaels, Ph.D., M.P.H., is an epidemiologist and a nationally recognized leader in the scientific community's efforts to protect the integrity of the science on which public health and regulatory policies are based. Before joining the Occupational Safety and Health Administration (OSHA), he was professor of environmental and occupational health at the GWU School of Public Health. From 1998 to 2001, Dr. Michaels served as assistant secretary of Energy for Environment, Safety and Health. In that position, he was the chief architect of the Energy Employees Occupational Illness Compensation Program, the historic initiative to compensate nuclear weapons workers who contracted occupational illnesses as a result of exposure to radiation, beryllium, and other hazards. The program has provided more than $6 billion in payments to sick workers and the families of deceased workers. In 2006, Dr. Michaels was awarded the American Association for the Advancement of Science's Scientific Freedom and Responsibility Award, and, in 2009, the John P. McGovern Science and Society Award given by Sigma Xi, the Scientific Research Society, for his work in scientific integrity and for gaining compensation for nuclear weapons workers. Dr. Michaels is the author of studies examining the health of construction workers, printers, bus drivers, and other occupations, as well as of numerous publications on science and regulatory policy. He is a graduate of the City College of New York, and holds an M.P.H. and Ph.D. from Columbia University.

Aubrey Miller, M.D., M.P.H., joined NIEHS to serve as senior medical advisor and NIEHS liaison to the U.S. Department of Health and Human Services (HHS). Miller's office is located on the NIH Campus in Bethesda where he oversees a small staff of NIEHS employees who are readily available to meet with NIH and HHS representatives, federal partners, members of Congress, and other stakeholders to discuss how environmental factors influence human health and disease. A medical epidemiologist and a captain in the U.S. Public Health Service, Dr. Miller has longstanding experience, publications, and contributions to a wide range of occupational and environmental health issues and policies. Dr. Miller previously served as the chief medical officer for the U.S. Food and Drug Administration Office of Counterterrorism and Emerging Threats. Previously, he worked as a senior medical officer and regional toxicologist for the EPA and for the HHS Office of the Secretary in Denver, providing leadership, expertise, and coordination for multiagency emergency responses, such as the Libby, Montana, asbestos situation, the anthrax attacks in Washington, DC, and Hurricane Katrina. He also conducted more than 30 field investigations while working for several years as a medical officer for CDC and NIOSH. Dr. Aubrey received his M.D. from Rush Medical College in Chicago and his M.P.H. in environmental and occupational health sciences from the University of Illinois School of Public Health. He is board certified in occupational and environmental medicine. He is a member of the American Public Health Association, American College of Occupational and Environmental Medicine, and American Conference of Governmental Industrial Hygienists.

Jennifer Orme-Zavaleta, Ph.D., has been with the EPA for 30 years, working in the areas of human health and ecological research, risk assessment, policy and regulation development, strategic planning, and program implementation. The focus of her experience includes the evaluation of risks to human and ecosystem health, and the influence of environmental change on human health in response to a variety of stressors, including synthetic organic and inorganic chemicals, radionuclides, microorganisms, and vector-borne disease. She has worked in the Offices of Research and Development, Pesticides and Toxic Substances, and Water. As interim national program director for Safe and Sustainable Water Resources, she led the realignment of the former drinking water and water quality research programs to form a holistic research program that maximizes responsiveness to the rapidly changing needs of the agency's water program, regional offices, and other critical water resource partners and stakeholders. During her career, she has been involved with the risk assessment practices within the agency, and the national and international scientific community. As a member of the EPA's Risk Assessment Forum Technical Panel she was one of several scientists who developed the Guideline for Reproductive Risk Assessment, Guideline for Implementation of EPA's Cancer Risk Assessment Guidelines, and Guideline for Assessing Risk from Less than Lifetime Exposures. She has also served as the manager of the EPA's Drinking Water Health Advisory Program, leading the development of more than 120 health advisories for inorganic, organic, pesticide, munition, and microbial contaminants. These assessments have been used by the World Health Organization to develop guidelines for drinking water quality and also serve as the basis for unreasonable risk to health determinations for U.S. public water supplies when regulatory violations occur.

Bob Perciasepe, M.Pl., M.P.A., returned to the EPA to serve as deputy administrator—the nation's second-ranking environmental official and the agency's chief operating officer—with his appointment by President Obama in 2009. In this role, he continues a career spanning nearly four decades as one of the nation's leading environmental and public policy figures. An expert on environmental stewardship, advocacy, public policy, and national resource and organizational management, Mr. Perciasepe is widely respected within both the environmental and U.S. business communities. His extensive experience includes service both inside and outside of government. He served as a top EPA official in the Clinton administration, appointed first as the nation's top water official and later as the senior official responsible for air quality across the United States. Prior to being named to his current position, he was chief operating officer at the National Audubon Society, one of the world's leading environmental organizations. He has also held top positions within state and municipal government, including as Secretary of the Environment for the State of Maryland and as a senior official for the City of Baltimore. Perciasepe holds a bachelor of science degree in natural resources from Cornell University and a master's degree in planning and public administration from the Maxwell School of Syracuse University.

Christopher J. Portier, Ph.D., joined CDC in 2010 as the director of the National Center for Environmental Health and Agency for Toxic Substances and Disease Registry. Dr. Portier came to CDC from NIEHS, where he was the senior advisor to the director and a principal investigator in environmental systems biology. Formerly, Dr. Portier was associate director of NIEHS, director of the Environmental Toxicology Program at NIEHS, and associate director of NTP. Dr. Portier is an internationally recognized expert in the design, analysis, and interpretation of environmental health data. His research efforts and interests include such diverse topics as cancer biology, risk assessment, climate change, bioinformatics, immunology, neurodevelopment, genetically modified foods, and genomics. From 2000 to 2006, he managed NTP and developed a strategic initiative that is internationally recognized for its innovation. He has contributed to the development of cancer risk assessment guidelines for national and international agencies and has either directed or contributed significantly to numerous risk assessments. He led the U.S. evaluation of electromagnetic fields by national and international scientists, which was the first comprehensive review in this field. Dr. Portier directed efforts of the U.S. government to develop a collaborative research agenda with Vietnam on the health effects of Agent Orange in that country. He has just directed a multiagency review of research needs for the health effects of climate change for the entire U.S. government. He has served as an advisor to the Finnish Academy of Sciences on the Centers of Excellence Research Program, as a member of World Health Organization/International Agency for Research on Cancer scientific committees, and as a reviewer for grants for the United States, the European Union, and many other grant-sponsoring organizations. Dr. Portier received his B.Sc. degree (1977) in mathematics (summa cum laude) and his M.S. (1979) and Ph.D. (1981) degrees in biostatistics. He has authored more than 150 peer-reviewed publications, 30 book chapters, and 40 technical reports. In the past 5 years, he has given more than 70 invited lectures, many of them at international meetings.

Allen Robinson, Ph.D., has conducted research examining the technical and policy issues related to energy and the environment. A current focus is fine particulate matter, from which 50,000 Americans are estimated to die prematurely each year and almost 70 million people in the United States are affected because they live in areas that violate the National

Ambient Air Quality Standard. Atmospheric particles also have a controlling influence on Earth's climate and degrade visibility. A major thrust of Dr. Robinson's research is characterizing fine-particle emissions from combustion systems such as diesel engines. Laboratory experiments using dilution samplers and a smog chamber have revealed a dynamic new picture for primary organic aerosol emissions, in which these emissions evaporate, oxidize, and recondense over time. These findings require updated approaches to measure and simulate emissions from combustion systems. His group is working to implement this revised framework into chemical transport models to investigate its implications on our understanding of urban, regional, and global air quality. This modeling has revealed a potentially important new source of regional oxidized and presumably hydrophilic organic aerosol. Work is ongoing to better understand the health consequences and climate effects of these pollutants. Dr. Robinson joined Carnegie Mellon in 1998 after working for 2 years as a postdoctoral fellow at the Combustion Research Facility at Sandia National Laboratories. He received his Ph.D. from UC Berkeley in mechanical engineering in 1996 and his B.S. in civil engineering from Stanford University in 1990. Dr. Robinson received the Ahrens Career Development Chair in Mechanical Engineering from Carnegie Mellon University in 2005 and the George Tallman Ladd Outstanding Young Faculty Award from Carnegie Mellon University in 2000.

Deborah Swackhamer, Ph.D., is an environmental chemist with an emphasis in aquatic chemistry. She manages the University of Minnesota Water Research Center's research and educational programs, including overseeing the Water Resources Research Institute grants program for the USGS and developing research and educational opportunities for the center. She is a professor of environmental chemistry in the University of Minnesota's School of Public Health and holds the Charles M. Denny Chair of Science, Technology, and Public Policy in the University's Humphrey Institute of Public Affairs. Dr. Swackhamer received her B.A. in chemistry from Grinnell College and her M.S. in water chemistry and Ph.D. in limnology and oceanography from the University of Wisconsin–Madison. Her research focuses on the chemical and biological processes that control the fate of toxic organic contaminants in the environment, environmental exposure, and risk assessment.

Aaron Wernham, M.D., is the director of the Health Impact Project, a collaboration of the Robert Wood Johnson Foundation and The Pew Charitable Trusts designed to promote the use of health impact assessments (HIAs) and support the growth of the field in the United States. Dr. Wernham is an HIA expert who has led HIAs at the state and federal levels. He has conducted HIA training for, collaborated with, and advised numerous health and environmental regulatory agencies on integrating HIAs into their programs. Prior to joining Pew, Dr. Wernham was a senior policy analyst with the Alaska Native Tribal Health Consortium, where he led the first successful efforts in the United States to formally integrate HIAs into the federal environmental impact statement process. Dr. Wernham also directed a collaborative state–tribal–federal working group on HIAs and, with this group, wrote HIA guidance for federal and state environmental regulatory and permitting efforts. Dr. Wernham received his medical degree from UC San Francisco, and a master's degree in health and medical sciences from UC Berkeley. Board certified in family medicine, he previously served as clinical faculty in the UC Davis family medicine residency program at Contra Costa Regional Medical Center.

Roxana Zulauf Witter, M.D., M.S.P.H., M.S., is an assistant research professor in the Department of Occupational and Environmental Health at the Colorado School of Public Health. She led a health impact assessment investigating potential health effects of natural gas development in a residential community in Colorado. She also led the development of a white paper and literature review of potential exposure-related human health effects of oil and gas development. Dr. Witter is co-program director of the Occupational and Environmental Medicine Residency and teaches the Occupational and Environmental Toxicology course at Colorado School of Public Health. Dr. Witter is board certified in occupational and environmental medicine and spent several years in clinical practice.

C

Acronyms

AMCV	air monitoring comparison value
ATSDR	Agency for Toxic Substance and Disease Registry
AutoGC	auto gas chromatography
bcfd	billion cubic feet per day
CAS	Chemical Abstract Services
CDC	Centers for Disease Control and Prevention
DFW	Dallas–Fort Worth
EIS	environmental impact statement
EPA	U.S. Environmental Protection Agency
GAO	U.S. Government Accountability Office
GWP	global warming potential
HAP	hazardous air pollutant
HI	hazard index
HIA	health impact assessment
HQ	hazard quotient
ICMM	International Council on Mining and Metals
IFC	International Finance Corporation
IOM	Institute of Medicine
kWh	kilowatt-hour
MCF	thousand cubic feet
MSETC	Marcellus Shale Education & Training Center

NCEH	National Center for Environmental Health
NIEHS	National Institute of Environmental Health Sciences
NIH	National Institutes of Health
NIOSH	National Institute for Occupational Safety and Health
NORM	naturally occurring radioactive material
NO_x	nitrogen oxides
NRC	National Research Council
OSHA	Occupational Safety and Health Administration
PM	particulate matter
ppb	parts per billion
RfC	reference concentration
SO_2	sulfur dioxide
STEPS	Service, Transmission, Exploration, and Production Safety Network
TCEQ	Texas Commission on Environmental Quality
USGS	U.S. Geological Survey
VMT	vehicle miles traveled
VOC	volatile organic compound